METHODS IN MOLECULAR BIOLOGY

For further volumes:
http://www.springer.com/series/7651

For over 35 years, biological scientists have come to rely on the research protocols and methodologies in the critically acclaimed *Methods in Molecular Biology* series. The series was the first to introduce the step-by-step protocols approach that has become the standard in all biomedical protocol publishing. Each protocol is provided in readily-reproducible step-by-step fashion, opening with an introductory overview, a list of the materials and reagents needed to complete . the experiment, and followed by a detailed procedure that is supported with a helpful notes section offering tips and tricks of the trade as well as troubleshooting advice. These hallmark features were introduced by series editor Dr. John Walker and constitute the key ingredient in ,each and every volume of the *Methods in Molecular Biology* series. Tested and trusted, comprehensive and reliable, all protocols from the series are indexed in PubMed.

Lysosomes

Methods and Protocols

Edited by

Stephanie M. Hughes and Indranil Basak

Department of Biochemistry, Faculty of Biomedical Sciences, Brain Health Research Centre, University of Otago, Dunedin, New Zealand

Humana Press

Editors
Stephanie M. Hughes
Department of Biochemistry
Faculty of Biomedical Sciences
Brain Health Research Centre
University of Otago
Dunedin, New Zealand

Indranil Basak
Department of Biochemistry
Faculty of Biomedical Sciences
Brain Health Research Centre
University of Otago
Dunedin, New Zealand

ISSN 1064-3745 ISSN 1940-6029 (electronic)
Methods in Molecular Biology
ISBN 978-1-0716-4843-8 ISBN 978-1-0716-4844-5 (eBook)
https://doi.org/10.1007/978-1-0716-4844-5

This Humana imprint is published by the registered company Springer Science+Business Media, LLC, part of Springer Nature.
The registered company address is: 1 New York Plaza, New York, NY 10004, U.S.A.

Preface

Lysosomes are multifunctional organelles central to cellular homeostasis, mediating cellular degradation and recycling. Beyond degradation, lysosomes are involved in immune responses, plasma membrane repair, cell death, and metabolic signalling and play crucial additional roles in neurons. Cellular homeostasis requires that the lysosome lumen is maintained at acidic pH (4.5–5.0), degradative enzymes are correctly trafficked to the lysosome, membrane proteins required for substrate import and metabolite export are functional, and the organelles are effectively trafficked within the cell.

Lysosomal dysfunction is implicated in many diseases, including lysosomal storage disorders (LSDs) where the primary defect lies a gene involved in lysosome function, most commonly a lysosomal enzyme. Dysregulated lysosome function is common in neurodegeneration, either by virtue of loss of lysosomal protein function in LSDs or likely a secondary consequence in sporadic neurodegenerative conditions such as Alzheimer's disease. Aberrant gain of function is associated with many cancers and infections.

Modern lysosome research leverages a wide range of complementary methods—from biochemical assays and advanced imaging to organelle isolation and disease modelling. In this book, we have gathered a range of commonly used and new methodologies. These chapters are aimed at distilling the essential methodologies and giving expert insight to key features with notes pertaining to key elements of the methods.

The book starts with fundamental analysis of lysosomes using enzyme assays, both in situ and from cell lysates (Chap. 1). In Chaps. 2 and 3, lysosome quantification and pH methods are described using either microscopy or flow cytometry, respectively. The latter compares relative and absolute pH assessment using either dextran derivates or Lysotracker probes. Photothermal imaging, a method to image lysosomes using their inherent heat produced when light energy is absorbed by endogenous dyes, is described together with the hardware setup required in Chap. 4. In Chaps. 5 and 6, methods to analyse autophagic flux, crucial for delivery of substrates to the lysosome, are described. Chapter 5 illustrates the use of western blotting and LC3 analysis while Chap. 6 illustrates a recent methodology to measure autophagy in human blood. From here we move to two methods to isolate and analyse lysosomes, in Chap. 7, superparamagnetic iron oxide nanoparticles (SPIONs) are used to isolate lysosomes. Chapter 8 describes lysoIP using TMEM192-HA labelled cells or the LysoTag mouse. Methods for downstream metabolite and proteome analysis are also described here and in Chap. 9.

The importance of membrane contact sites (MCS) in transfer of lipids and ions between organelles has been recently established. In Chap. 10, transmission electron microscopy methods are used to analyse lysosome-mitochondria MCS. Transport of lysosomes within cells relies on microtubules and is especially important for transport within neuronal axons and dendrites. Chapter 11 described a method of analysing lysosome trafficking in human iPSC-neurons.

Model systems are crucial to understand lysosome homeostasis and dysfunction in disease. In the next chapters, methods to generate and characterise various models are described. Human iPSC lines derived from either patient cells or generated by CRISPR editing are valuable model systems. In Chap. 12, the derivation of LSD iPSC lines is described. To study lysosome function and disease in whole organisms, we offer protocols in yeast, *Schizosaccharomyces pombe* (Chap. 13), the slime mould, *Dictystelium discoideum* (Chap. 14), and zebrafish (Chap. 15), followed by immunohistochemical analysis of lysosomal markers in various mouse tissues (Chap. 16).

We thank all the authors for their valuable contributions and hope this is a useful resource for those new to the field as well as those looking to expand their technique repertoire. We also thank Prof John Walker for his assistance in the early phases of this project.

Dunedin, New Zealand

Stephanie M. Hughes
Indranil Basak

Contents

Contributors

MONTHER ABU-REMAILEH • *Department of Chemical Engineering, Stanford University, Stanford, CA, USA; Department of Genetics, Stanford University, Stanford, CA, USA; The Institute for Chemistry, Engineering and Medicine for Human Health (Sarafan ChEM-H), Stanford University, Stanford, CA, USA; The Phil & Penny Knight Initiative for Brain Resilience at the Wu Tsai Neurosciences Institute, Stanford University, Stanford, CA, USA*

INDRANIL BASAK • *Department of Biochemistry, Faculty of Biomedical Sciences, Brain Health Research Centre, University of Otago, Dunedin, New Zealand*

JULIEN BENSALEM • *Lysosomal Health in Ageing, Hopwood Centre for Neurobiology, Lifelong Health Theme, South Australian Health and Medical Research Institute, Adelaide, SA, Australia; Faculty of Health and Medical Sciences, The University of Adelaide, Adelaide, SA, Australia*

HANNAH BEST • *Medicines Discovery Institute, Cardiff University, Cardiff, UK*

SARA BONINI • *Institute for Biochemistry and Molecular Biology, University of Bonn, Bonn, Germany*

SUEANNE CHEAR • *Wicking Dementia Research and Education Centre, University of Tasmania, Hobart, Australia*

ADELENE CHIAM • *Wicking Dementia Research and Education Centre, University of Tasmania, Hobart, Australia*

JOSÉ ÁNGEL CLEMENTE-RAMOS • *UCL Great Ormond Street Institute of Child Health, University College London, London, UK*

ANTHONY L. COOK • *Wicking Dementia Research and Education Centre, University of Tasmania, Hobart, Australia*

JONATHAN D. COOPER • *Department of Pediatrics, Washington University in St. Louis, School of Medicine, St. Louis, MO, USA; Department of Genetics, Washington University in St. Louis, School of Medicine, St. Louis, MO, USA; Department of Neurology, Washington University in St. Louis, School of Medicine, St. Louis, MO, USA*

OLUWATOBI N. EBODA • *Department of Biochemistry, Brain Health Research Centre, University of Otago, Dunedin, New Zealand*

CARLOS ENRICH • *Departament de Biomedicina, Unitat de Biologia Cel·lular, Facultat de Medicina i Ciències de la Salut, Centre de Recerca Biomèdica CELLEX, Institut d'Investigacions Biomèdiques August Pi i Sunyer (IDIBAPS), Universitat de Barcelona, Barcelona, Spain*

SIAN GARDINER • *Medicines Discovery Institute, Cardiff University, Cardiff, UK*

KATHRYN J. HATTERSLEY • *Lysosomal Health in Ageing, Hopwood Centre for Neurobiology, Lifelong Health Theme, South Australian Health and Medical Research Institute, Adelaide, SA, Australia*

LEANNE K. HEIN • *Lysosomal Health in Ageing, Hopwood Centre for Neurobiology, Lifelong Health Theme, South Australian Health and Medical Research Institute, Adelaide, SA, Australia*

ALEX W. HEWITT • *Menzies Institute for Medical Research, School of Medicine, University of Tasmania, Hobart, Australia*

ROBERT J. HUBER • *Environmental and Life Sciences Graduate Program, Trent University, Peterborough, ON, Canada; Department of Biology, Trent University, Peterborough, ON, Canada*

STEPHANIE M. HUGHES • *Department of Biochemistry, Faculty of Biomedical Sciences, Brain Health Research Centre, University of Otago, Dunedin, New Zealand*

WILLIAM D. KIM • *Environmental and Life Sciences Graduate Program, Trent University, Peterborough, ON, Canada*

EMILY KIRKHAM • *School of Biosciences, Cardiff University, Cardiff, UK*

NOUF N. LAQTOM • *Biological and Environmental Sciences and Engineering Division, King Abdullah University of Science and Technology (KAUST), Thuwal, Saudi Arabia*

STEPHANE LEFRANCOIS • *Centre Armand-Frappier Santé Biotechnologie, Institute national de la recherche scientifique, Laval, QC, Canada; Department of Anatomy and Cell Biology, McGill University, Montreal, QC, Canada; Centre d'Excellence en Recherche sur les Maladies Orphelines – Fondation Courtois (CERMO-FC), Université du Québec à Montré al (UQAM), Montreal, QC, Canada*

EMYR LLOYD-EVANS • *School of Biosciences, Cardiff University, Cardiff, UK; Medicines Discovery Institute, Cardiff University, Cardiff, UK*

VALENTINA MARCHICA • *Royal Veterinary College, London, UK*

KIRSTIN O. MCDONALD • *Department of Biochemistry, Faculty of Biomedical Sciences, Brain Health Research Centre, University of Otago, Dunedin, New Zealand*

JUN MIYAZAKI • *Faculty of Systems Engineering, Wakayama University, Wakayama, Japan*

SARA E. MOLE • *UCL Great Ormond Street Institute of Child Health, University College London, London, UK*

PATHMA MUTHUKOTTIAPPAN • *Institute for Biochemistry and Molecular Biology, University of Bonn, Bonn, Germany*

HEMANTH R. NELVAGAL • *Department of Pediatrics, Washington University in St. Louis, School of Medicine, St. Louis, MO, USA*

JENNIFER E. PALMER • *Department of Medical Genetics, Cambridge Institute for Medical Research, Cambridge, UK; UK Dementia Research Institute, Cambridge Institute for Medical Research, University of Cambridge, Cambridge, UK*

CLAUDIA PARRA-RUIZ • *Departamento de Gastroenterología, Facultad de Medicina, Pontificia Universidad Católica de Chile, Santiago, Chile*

NASTASSJA RAMDEEN • *Royal Veterinary College, London, UK*

DAVID C. RUBINSZTEIN • *Department of Medical Genetics, Cambridge Institute for Medical Research, Cambridge, UK; UK Dementia Research Institute, Cambridge Institute for Medical Research, University of Cambridge, Cambridge, UK*

CLAIRE RUSSELL • *Royal Veterinary College, London, UK*

TIMOTHY J. SARGEANT • *Lysosomal Health in Ageing, Hopwood Centre for Neurobiology, Lifelong Health Theme, South Australian Health and Medical Research Institute, Adelaide, SA, Australia; Faculty of Health and Medical Sciences, The University of Adelaide, Adelaide, SA, Australia*

ETIENNE SAUVAGEAU • *Centre Armand-Frappier Santé Biotechnologie, Institute national de la recherche scientifique, Laval, QC, Canada*

BIGGY SIMBI • *Royal Veterinary College, London, UK*

JANA TALBOT • *Wicking Dementia Research and Education Centre, University of Tasmania, Hobart, Australia*

BRYONY N. THORNE • *Wicking Dementia Research and Education Centre, University of Tasmania, Hobart, Australia*

RYOTARO WAGATSUMA • *Faculty of Systems Engineering, Wakayama University, Wakayama, Japan*
HELEN WALLER-EVANS • *Medicines Discovery Institute, Cardiff University, Cardiff, UK*
SOPHIE H. WANG • *Department of Pediatrics, Washington University in St. Louis, School of Medicine, St. Louis, MO, USA*
EMMA J. WILKINSON • *Wicking Dementia Research and Education Centre, University of Tasmania, Hobart, Australia*
DOMINIC WINTER • *Department of Metabolism, Senescence and Autophagy, Research Center One Health Ruhr, University Alliance Ruhr, University Hospital Essen, University of Duisburg-Essen, Essen, Germany; Institute for Biochemistry and Molecular Biology, University of Bonn, Bonn, Germany*
EMILY MALINI YOUNG • *Royal Veterinary College, London, UK*
SILVANA ZANLUNGO • *Departamento de Gastroenterología, Facultad de Medicina, Pontificia Universidad Católica de Chile, Santiago, Chile*
EWA A. ZIÓŁKOWSKA • *Department of Pediatrics, Washington University in St. Louis, School of Medicine, St. Louis, MO, USA*

Chapter 1

Determining Lysosomal Enzyme Activity Using Fluorogenic Probes

Etienne Sauvageau and Stephane Lefrancois

Abstract

Lysosomes are responsible for a number of cellular functions, including the degradation of various biological molecules. Soluble enzymes within the lysosomal lumen are required to perform this function. Lysosomal activity can be disrupted in a variety of diseases, and measuring the activity of specific enzymes can be performed. In this chapter, we detail how lysosomal enzyme activity can be measured either in cell lysates or intact cells. This can be used to study fundamental cell biology or the effect of therapeutics targeting lysosomal function.

Key words Lysosomal enzyme activity, Fluorogenic substrates, Batten disease, Lysosomes

1 Introduction

Lysosomes are membrane-bound organelles that serve various functions in the cell, including autophagy, nutrient sensing, and cellular metabolism [1, 2]. However, they are better known and were originally identified for their function in cellular waste management and disposal/recycling. Due to their content that includes a variety of enzymes, lysosomes can degrade several biological molecules, including proteins, carbohydrates, and fatty acids [1].

Lysosomes are composed of integral membrane proteins such as lysosomal-associated membrane protein 1 (LAMP-1) and LAMP-2, which are thought to regulate lysosomal structural integrity and pH. Lysosomal integral membrane protein 2 (LIMP-2) is also found in the membrane of lysosomes and is thought to play a role in lysosome biogenesis and function [3].

Lysosomes also contain soluble proteins, which include enzymes such as cathepsins and nonenzymatic proteins such as prosaposin. These proteins are required for the digestion of biological molecules. Soluble lysosomal proteins are transported from the trans-Golgi Network (TGN) to endosomes, to eventually

Stephanie M. Hughes and Indranil Basak (eds.), *Lysosomes: Methods and Protocols*, Methods in Molecular Biology, vol. 2976, https://doi.org/10.1007/978-1-0716-4844-5_1,

localize to lysosomes. In order for them to be sorted, these proteins interact with one of three lysosomal sorting receptors in the TGN, the cationic independent mannose 6-phosphate receptor (CI-MPR), cationic dependent mannose 6-phosphate receptor (CD-MPR), and sortilin [4, 5]. Soluble lysosomal proteins can interact with one or several of the receptors [6]. The receptor/cargo complex is then packaged into clathrin-coated vesicles, which bud from the TGN, travel to endosomes, and fuse with this compartment to deliver cargo [7, 8]. The soluble proteins then move onto lysosomes, while the receptors are retrieved back to the TGN for subsequent rounds of trafficking [9–11].

Mutations in a variety of soluble lysosomal proteins can cause lysosomal diseases, which generally result in the accumulation of a nondegraded substrate. For example, mutations in β-glucocerebrosidase cause the accumulation of glucocerebroside, resulting in Gaucher's disease. Disruption of the trafficking pathway can also result in human disease. For example, mutations in ceroid lipofuscinosis-3 (CLN3), the protein mutated in CLN3 Batten disease, result in inefficient retrieval of the lysosomal sorting receptors [12, 13]. This results in lysosomes that accumulate several substrates, including subunit C of mitochondrial ATP synthase (SCMAS). Lysosomal dysfunction is also present in more prominent diseases, such as Parkinson's disease (PD). Retromer is one of the protein complexes that participates in the retrieval of the lysosomal sorting receptors, and mutations in one of its subunits have been linked to PD [14, 15].

To monitor lysosomal enzyme activity, cells or cell lysates can be provided with fluorogenic substrates that can be cleaved by specific enzymes. When these compounds are intact, the fluorescence is quenched. Upon cleavage in lysosomes, the fluorescence signal is released. A stronger fluorescence signal is interpreted as a more active enzyme. Using this system, one can determine the activity of several enzymes in various contexts.

It is clear that monitoring lysosomal function is necessary in order to understand fundamental lysosomal biology and to understand how disease can affect its function. Beyond this, studying lysosomal function will also allow us to determine if treatment of these diseases is working. In this chapter, we will provide information on how to study lysosomal activity in live cells and from cell lysates.

2 Materials

2.1 Cell Culture Materials

2.1.1 Cell Lines

1. HEK293 and HeLa were originally obtained from ATCC (Manassas, VA) and maintained in Dulbecco's modified Eagle's medium (DMEM) supplemented with 10% FBS with regular testing for contamination.

2.1.2 Reagents

1. Polyethylenimine, linear MW 25,000 Da (PEI) (Polysciences, Inc., Warrington, PA).
2. 96-well microplate, black walled, non-binding (655900, Greiner Bio-One, Kremsmünster, Austria).
3. Bio-Rad Protein Assay Dye Reagent Concentrate (5000006, Bio-Rad Laboratories, Mississauga, ON).

2.1.3 Sterile PBS

1. To prepare 10× PBS, dissolve 8 g of NaCl, 0.2 g of KCl, 1.44 g of Na_2HPO_4, and 0.24 g of KH_2PO_4 in 800 mL of ddH_2O. Use HCl to adjust the pH to 7.4, then add more ddH_2O to reach 1000 mL. To prepare 1×, add 100 mL of 10× PBS to 900 mL of ddH_2O. Sterilize PBS by autoclaving and store at room temperature.

2.1.4 PBS-EDTA 5 mM

1. Add 1 mL of EDTA 0.5 M to 99 mL of PBS.

2.1.5 0.05% Trypsin

1. This reagent can be purchased. Once received, thaw the stock and aliquot into a 15-mL tube. Freeze until ready to use.

2.1.6 DMEM Supplemented with 10% FBS

1. Purchased DMEM can be supplemented with FBS. Add 50 mL of FBS to 450 mL of DMEM and pass through a 0.22-μm polyethersulfone (PES) filter.

2.1.7 Lysis Buffers

TNE

1. (50 mM Tris–HCl, 150 mM NaCl, 2 mM EDTA, 1% Triton-X-100)
2. In a 50-mL tube, add 40 mL of ddH_20, 2.5 mL of Tris–HCl 1 M pH 7.5, 1.875 mL of NaCl 4 M, 200 μL of 0.5 M EDTA, and 500 μL of Triton-X-100. Gently agitate until Triton-X-100 is fully dissolved and complete to 50 mL with ddH_20.

2.1.8 Radioimmunoprecipitation Assay Buffer (RIPA)

1. (50 mM Tris–HCl pH 8.0, 150 mM NaCl, 2 mM EDTA, 1% IGEPAL, 0.5% sodium deoxycholate, 0.1% SDS) (*see* **Note 1**)
2. Combine 5 g of sodium deoxycholate in 30 mL of ddH_2O and heat to 50 °C to dissolve, then let cool to RT. Store at RT and protect from light. Combine 3 mL of 5 M NaCl, 5 mL of 1 M Tris–HCL, pH 8.0, 1 mL of IGEPAL, and 1 mL of 10% SDS. Complete to 100 mL with ddH_2O.

2.2 Preparation of Substrate

Dissolve substrate in DMSO, or according to the manufacturer's instructions, to obtain a 1000× or 50× stock solution the day of the experiment (*see* **Note 2**). If this is impractical due to the amount of substrate supplied, prepare a 1000× stock solution, prepare aliquots of 10 or 20 μL, and place in a −20 °C freezer to store.

3 Methods

We generally perform these types of experiments using cell lines. Whenever possible, we prefer to perform these experiments in intact cells. If working with intact cells is not feasible, or if the substrate is not cell permeable (*see* **Note 3**), perform these experiments using cell lysates. It is also possible to perform these experiments with tissue lysates.

3.1 Plating Cells

1. Plate 2×10^5 HEK293 or HeLa cells in 2-cm plates (6-well plates) with 2 mL of DMEM supplemented with 10% FBS. The goal is to obtain cells at 50% confluence the next day (*see* **Note 4**). Make sure to include cells that will not be treated to serve as a control for Subheading 3.2. This is not necessary for Subheading 3.3.
2. If required, transfect cells 24 h after plating according to your protocol. We generally use polyethylenimine, linear MW 25,000 Da (PEI), but if trying to rescue cells, a lipid-based system is better as the percentage of cells expressing the exogenous protein is increased.
3. The experiment is generally performed 48 h after transfection, but other time points can be used. If not transfecting the cells, the experiment can be performed 24 or 48 h after plating the cells.
4. When ready to proceed with the experiment, a decision needs to be made whether to perform the experiment on live cells (Subheading 3.2) or in cell lysates (Subheading 3.3) (*see* **Note 3**).

3.2 Experiments in Live Cells

1. Warm DMEM supplemented with 10% FBS to 37 °C.
2. Prepare a 1× concentration of substrate in DMEM supplemented with 10% FBS.
3. Replace the cell culture media with substrate-supplemented media from **step 2**. As a control to determine the background fluorescence of the cells, include cells that have not been supplemented with substrate.
4. Incubate cells with substrate for the desired time (*see* **Note 5**).
5. Wash cells once with 1 mL of PBS, then add 1 mL of PBS containing EDTA and incubate for 5 min at 37 °C (*see* **Note 6**).
6. Transfer cells with media to a cuvette to measure the OD 600 on a spectrophotometer to quantify the cells (*see* **Note 7**).
7. Transfer cells to a 1.5-mL tube, and centrifuge at 800 g for 3 min to collect the cells.

8. Aspirate supernatant and resuspend the cell pellet in 500 μL of PBS. Adjust volume based on OD 600 so that the density of cells is equivalent in each condition.
9. Transfer 100 μL of each condition, in triplicate, to a black-walled, opaque-bottom 96-well plate.
10. Read fluorescence using a microplate reader using the excitation/emission spectrum of the substrate. These are generally in the 320 (for low pH in the range of 1.97–6.72) and 360 (for high pH in the range of 7.12–10.3) for excitation, while emission is generally in the 445–455 nm range (*see* **Note 8**).
11. To analyze the results, subtract the fluorescence reading obtained from cells with no substrate, from the reading of the cells with substrate (Fig. 1).

3.3 Experiment on Cell Lysate

1. Nonidet P-40 has been replaced by IPEGAL.
2. Wash cells once with 1 mL of PBS, then add 1 mL of PBS containing EDTA, and incubate for 5 min at 37 °C (*see* **Note 4**). Transfer cells to 1.5-mL tubes.
3. Add 150 μL of lysis buffer (TNE or RIPA) and incubate for 60 min at 4 °C with rotation (*see* **Note 9**).
4. Centrifuge tubes at 4 °C for 15 min at 10000 g to clear cellular debris.
5. Transfer supernatant to a clean 1.5-mL tube, keep on ice.
6. Quantify protein concentration using a colorimetric protein assay. Add 798 μL of ddH$_2$O to a cuvette, then add 2 μL of lysate, followed by 200 μL of Protein Assay Dye Reagent. Mix using an up and down motion using a 1000-μL pipette and incubate for 5 min. Read the OD at 595 on a spectrophotometer (*see* **Note 10**).
7. Prepare a 50×concentration of substrate in DMSO, then add PBS to achieve a 5× concentration.
8. Add 80 μL of PBS to 4 wells of a black-walled, opaque-bottom 96-well plate to serve as a control.
9. Add 25 μg of protein lysate to 4 other wells. Volume will vary based on the protein concentration as determined from **step 5** above, and complete with PBS to reach a total volume of 80 μL (*see* **Note 11**).
10. Add 20 μL of 5× substrate solution to each well, so that each well has a total volume of 100 μL.
11. Incubate the plate at 37 °C and read the plate at desired time points (*see* **Note 12**).
12. To analyze the result, subtract the fluorescence reading taken from the control that only contained substrate from the other readings containing both substrate and cell lysate (Fig. 2).

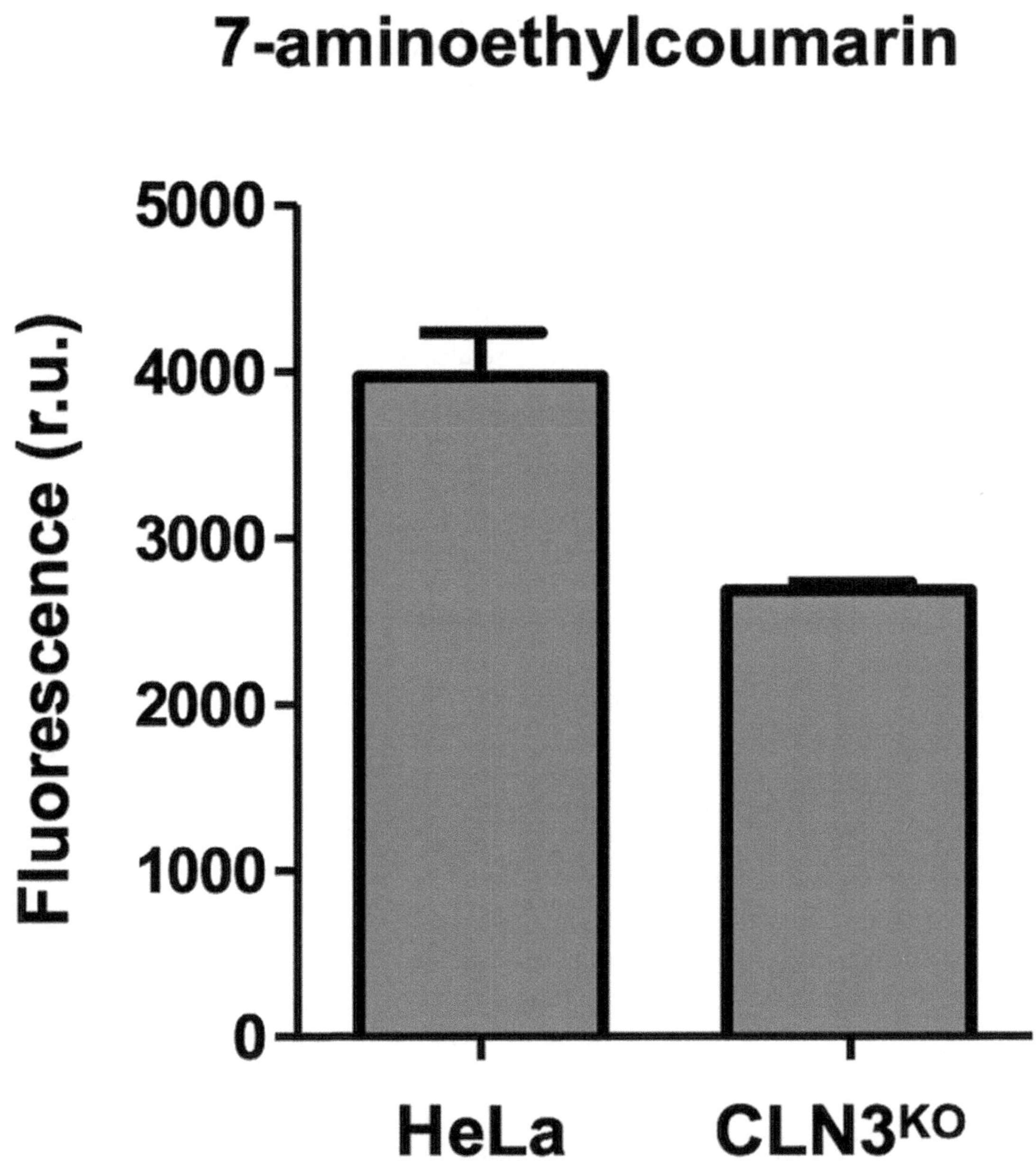

Fig. 1 Wild-type and CLN3^{KO} HeLa cells were incubated with CLN2 substrate 7-amino-4-methylcoumarin. Graphs show decreased activity in cells lacking CLN3

4 Notes

1. IGEPAL has replaced nonidet P40.
2. The information to dissolve the compounds can be found on the Cayman Chemical Website (https://www.caymanchem.com). While a 50× stock solution is used in the cell lysate

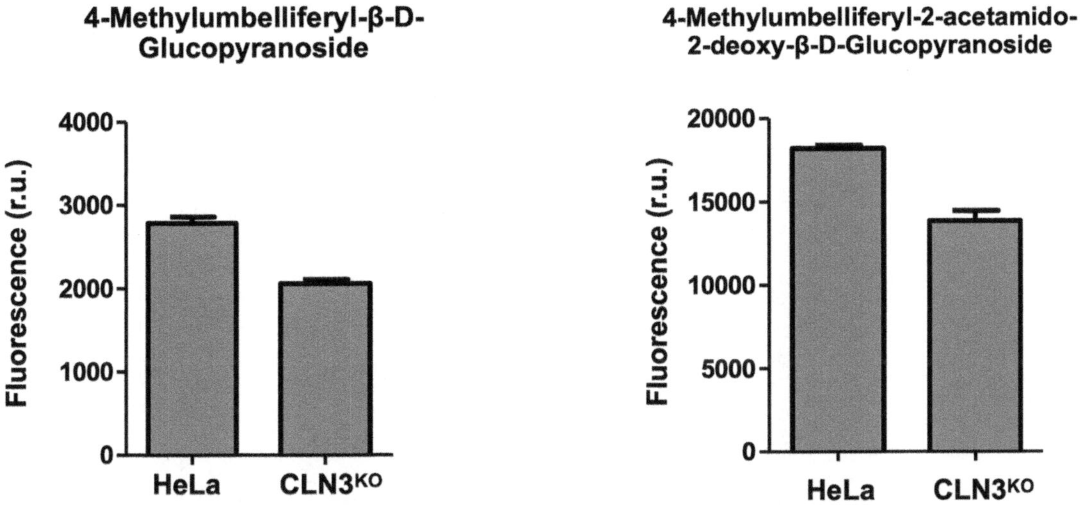

Fig. 2 Wild-type and CLN3^{KO} HeLa cells were incubated with the β-glucosidase substrate 4-methylumbelliferyl-α-D-glucopyranoside or the β-hexosaminidase substrate 4-methylumbelliferyl-2-acetamido-2-deoxy-β-D-glucopyranoside. Graphs show decreased activity in lysates lacking CLN3

experiments, a 1000× stock solution is used in the intact cell experiments.

3. The experiment on live cells requires that the substrate be cell permeable. If the substrate is, we recommend this as the better choice, as lysosomes will still be intact. *See* Table 1 for details.
4. Cell numbers can vary between cell types.
5. We generally incubate cells with the substrate for 15 min to 2 h. This will vary depending on the cell type, substrate, and plate reader. We suggest that several time points be measured when first performing these experiments to determine the optimal incubation time for each condition.
6. The incubation time in PBS-EDTA can vary depending on the cell type. This is for HeLa cells, while HEK293 cells will detach sooner.
7. OD measurement provides a faster method to determine cell concentration, and in our hands, provides consistent results comparable to cell counting.
8. Both TNE and RIPA work for these assays in our hands. We have not noticed any discernible differences in performing these experiments with either buffer.
9. Although optimal enzyme performance should occur at lower pH, which would more closely resemble the pH of intact lysosomes, we find that this results in quenching of the fluorescence. Performing these experiments at a more neutral pH is a trade-off between quenching and optimal lysosomal enzyme activity.

Table 1
Substrates used to measure lysosomal enzyme activity

Enzyme	Substrate	Final assay concentration	Incubation time	Excitation (pH 7.4)	Emission	Cell permeable
β-glucosidase	4-Methylumbelliferyl-β-D-glucopyranoside	5 mM	60 min	370	445–454	Not tested
α-glucosidase	4-Methylumbelliferyl-α-D-glucopyranoside	2 mM	30 min	360	445–454	Not tested
α-galactosidase	4-Methylumbelliferyl-α-D-galactopyranoside	5 mM	120 min	370	445–454	Not tested
β-hexosaminidase	4-Methylumbelliferyl-2-acetamido-2-deoxy-β-D-glucopyranoside	5 mM	15–30 min	360	445–455	NO
CLN1 (Palmitoyl protein thioestarase 1)	4-Methylumbelliferyl 6-thio-palmitate-β-D-glucopyranoside	5 mM	120 min	360	445–455	YES
N-acetylgalactosamine-6-sulphatase (GALNS)	4-Methylumbelliferyl β-D-galactopyranoside-6-sulfate (sodium salt)	20 mM	120 min	360	445–455	Not tested
β-glucuronidase	4-Methylumbelliferyl-β-D-glucuronide (hydrate)	5 mM	120 min	360	445–455	NO
CLN2 (tripeptidyl peptidase 1)	Ala-Ala-Phe-7-amino-4-methylcoumarin	1 mM	120 min	345	445	YES

10. The amount of lysate suggested is based on the BSA standard curve that we have developed in our lab. The amount might need to be adjusted depending on the curve used in your own lab.
11. We find that adding PBS to the well prior to the protein lysate is best to prevent protein degradation while pipetting.
12. Some substrates will have a good reading after 15 min, while others may take 1 or 2 h.

Acknowledgments

We thank Dr. Guido Hermey (Institute for Molecular and Cellular Cognition, Center for Molecular Neurobiology Hamburg, University Medical Center Hamburg-Eppendorf, 20251 Hamburg, Germany) for providing the parental and CLN3 knockout (CLN3^{KO}) HeLa cells. Funding for this work was provided by the Canadian Institutes of Health Research (PJT-173419) and the ForeBatten Foundation. We thank Icten Meras for helpful discussions and critical reading of this work.

References

1. Ballabio A, Bonifacino JS (2020) Lysosomes as dynamic regulators of cell and organismal homeostasis. Nat Rev Mol Cell Biol 21(2): 101–118. https://doi.org/10.1038/s41580-019-0185-4
2. Settembre C, Perera RM (2024) Lysosomes as coordinators of cellular catabolism, metabolic signalling and organ physiology. Nat Rev Mol Cell Biol 25(3):223–245. https://doi.org/10.1038/s41580-023-00676-x
3. Gonzalez A, Valeiras M, Sidransky E, Tayebi N (2014) Lysosomal integral membrane protein-2: a new player in lysosome-related pathology. Mol Genet Metab 111(2):84–91. https://doi.org/10.1016/j.ymgme.2013.12.005
4. Griffiths G, Hoflack B, Simons K, Mellman I, Kornfeld S (1988) The mannose 6-phosphate receptor and the biogenesis of lysosomes. Cell 52(3):329–341
5. Lefrancois S, Zeng J, Hassan AJ, Canuel M, Morales CR (2003) The lysosomal trafficking of sphingolipid activator proteins (SAPs) is mediated by sortilin. Embo J 22(24): 6430–6437
6. Ni X, Morales CR (2006) The lysosomal trafficking of acid sphingomyelinase is mediated by sortilin and mannose 6-phosphate receptor. Traffic 7(7):889–902
7. Puertollano R, Aguilar RC, Gorshkova I, Crouch RJ, Bonifacino JS (2001) Sorting of mannose 6-phosphate receptors mediated by the GGAs. Science 292(5522):1712–1716
8. Canuel M, Lefrancois S, Zeng J, Morales CR (2008) AP-1 and retromer play opposite roles in the trafficking of sortilin between the Golgi apparatus and the lysosomes. Biochem Biophys Res Commun 366(3):724–730
9. Arighi CN, Hartnell LM, Aguilar RC, Haft CR, Bonifacino JS (2004) Role of the mammalian retromer in sorting of the cation-independent mannose 6-phosphate receptor. J Cell Biol 165(1):123–133
10. Seaman MN (2004) Cargo-selective endosomal sorting for retrieval to the Golgi requires retromer. J Cell Biol 165(1):111–122
11. Simonetti B, Danson CM, Heesom KJ, Cullen PJ (2017) Sequence-dependent cargo recognition by SNX-BARs mediates retromer-independent transport of CI-MPR. J Cell Biol 216(11):3695–3712. https://doi.org/10.1083/jcb.201703015
12. Yasa S, Modica G, Sauvageau E, Kaleem A, Hermey G, Lefrancois S (2020) CLN3

regulates endosomal function by modulating Rab7A-effector interactions. J Cell Sci 133(6). https://doi.org/10.1242/jcs.234047

13. Calcagni A, Staiano L, Zampelli N, Minopoli N, Herz NJ, Di Tullio G, Huynh T, Monfregola J, Esposito A, Cirillo C, Bajic A, Zahabiyon M, Curnock R, Polishchuk E, Parkitny L, Medina DL, Pastore N, Cullen PJ, Parenti G, De Matteis MA, Grumati P, Ballabio A (2023) Loss of the batten disease protein CLN3 leads to mis-trafficking of M6PR and defective autophagic-lysosomal reformation. Nat Commun 14(1):3911. https://doi.org/10.1038/s41467-023-39643-7
14. Follett J, Norwood SJ, Hamilton NA, Mohan M, Kovtun O, Tay S, Zhe Y, Wood SA, Mellick GD, Silburn PA, Collins BM, Bugarcic A, Teasdale RD (2014) The Vps35 D620N mutation linked to Parkinson's disease disrupts the cargo sorting function of retromer. Traffic 15(2):230–244. https://doi.org/10.1111/tra.12136
15. Follett J, Bugarcic A, Yang Z, Ariotti N, Norwood SJ, Collins BM, Parton RG, Teasdale RD (2016) Parkinson Disease-linked Vps35 R524W Mutation Impairs the Endosomal Association of Retromer and Induces alpha-Synuclein Aggregation. J Biol Chem 291(35):18283–18298. https://doi.org/10.1074/jbc.M115.703157

Chapter 2

Lysosomal pH Measurements via Dual-Imaging Microscopy

Emily Kirkham, Sian Gardiner, Emyr Lloyd-Evans, Hannah Best, and Helen Waller-Evans

Abstract

Dual-wavelength ratiometric fluorescent probes are the current gold standard for generating a quantitative measure of subcellular compartment ion contents. This chapter describes the use of fluorescent probes to measure lysosomal pH, encompassing probe selection, calibration, instrumentation, and the miniaturization of assays to use for high-throughput readouts.

Key words Lysosome, pH, Fluorescent microscopy, High-throughput, Dual-wavelength ratio imaging, Fluorescein isothiocyanate, LysoSensor

1 Introduction

The lysosome is an organelle with a diverse array of cellular functions. Once considered simply the "waste-disposal" center of the cell, lysosomes are now viewed as key regulators of cellular homeostasis through roles in signaling and nutrient sensing [1]. The lysosome receives substrates from both the endocytic and autophagic pathways, where the acidic environment of the lysosomal lumen allows for the degradation and recycling of both intra- and extracellular material [2]. This unique acidic environment (pH 4.5–5.0) is optimal for the maturation and activity of most lysosomal hydrolytic enzymes [3]. Altered lysosomal pH drastically perturbs function and is frequently implicated in aging and disease [4, 5].

The lysosome maintains its acidic environment largely through the action of the vacuolar-type ATPase (v-ATPase), a multimeric pump that utilizes ATP to pump protons into the lysosomal lumen, resulting in a 500–1000-fold H^+ gradient across the membrane [6]. The proton gradient generated via v-ATPase alters the transmembrane potential through the continuous build-up of positive charge, which in turn inhibits further v-ATPase pumping. To dissipate this net charge, a counterion transporter must either transport

Stephanie M. Hughes and Indranil Basak (eds.), *Lysosomes: Methods and Protocols*, Methods in Molecular Biology, vol. 2976, https://doi.org/10.1007/978-1-0716-4844-5_2,

cytoplasmic anions into the lumen, luminal cations to the cytoplasm, or a combination of both [7–9]. Furthermore, lysosomes continuously "leak" H^+, meaning a tight relationship of the v-ATPase with other transporters is required to maintain the electrochemical gradient [10].

Influx of Cl^-, the most abundant lysosomal lumen anion, into the lysosomal lumen has been proposed as the major driving force in providing a counterion to support the continued action of v-ATPase [11]. The CFTR Cl^- channel and the CLC-7 Cl^-/H^+ exchanger have both been proposed as mechanisms of Cl^- influx, with mutations in both transporters previously reported to alter lysosomal pH [12–14]. However, the exact role of Cl^- remains controversial, with more recent studies showing no pH deficit in cells lacking CFTR or CLC-7 and indicating that cytosolic Cl^- transport can support acidification—but is dispensable—and instead plays a role in the degradation of material through hydrolase activation [11, 15, 16]. Other proposed counter flux mechanisms include cation efflux, such as Ca^{2+} through TPCs or TRPML1 [17–19]. A more recent addition to proteins known to be involved in lysosomal pH regulation is the H^+/K^+ transporter, and Parkinson's disease risk gene, TMEM175 [20]. This is a lysosomal membrane protein capable of transporting both K^+ and H^+ ions in a pH-dependent manner. Under physiological lysosomal conditions with an acidic lumen, TMEM175 acts primarily as an H^+ channel, as pH-induced structural changes restrict the size of the pore, preventing passage of K^+ ions [21]. TMEM175 is additionally regulated by direct interaction with the lysosomal membrane proteins LAMP1 and LAMP2, which reduces H^+ efflux [22]. Loss of TMEM175 causes lysosomal hyperacidification, indicating a physiological role in maintaining optimal lysosomal pH [23]. While a complete understanding of the mechanistic basis of acidification remains to be determined, it is clear that disruption of lysosomal pH has hugely detrimental effects on the cell and has been implicated in a host of diseases—many of which are neurological—including Alzheimer's disease, Down syndrome, Wolfram syndrome, Parkinson's disease, Kufor–Rakeb syndrome, and frontotemporal lobar degeneration [23–28].

1.1 Background of Lysosomal pH Quantification Experiments

The ability to quantitatively measure lysosomal pH in situ is critical to our understanding of lysosomal function and disease. Pioneering studies by de Duve and colleagues observed the sequestration and concentration of weak bases to demonstrate the acidic nature of the lysosome [29]. This "ion-trapping" principle forms the basis of many lysosomal probes where weak bases, which are membrane permeable at a neutral pH, passively diffuse into acidic compartments where they become protonated and selectively accumulate [30]. Ion-trapping is the mechanistic basis for many commonly used lysosomal probes, including acridine orange, neutral red, *N*-(3-((2,4-dinitrophenyl)amino)propyl)-*N*-(3-aminopropyl)

Table 1
A summary of commonly used commercially available probes for investigating lysosomal pH

Probe	Schematic (acidic–alkaline)	pK_a	pH optima	Excitation/emission
Single-wavelength probes				
Acridine orange		~9.7		Red: 488/650–710 Green (DNA): 488/520 nm
DAMP	Secondary antibody required	N/A	N/A	N/A
LysoSensor blue DND-167	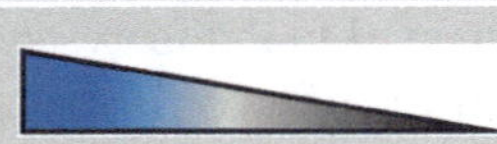	5.1	4.5–6.0	373/425 nm
LysoSensor green DND-189	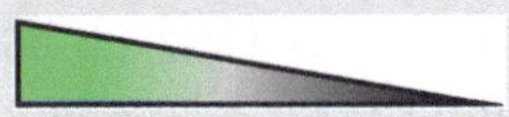	5.2	4.5–6.0	443/505 nm
Dual emission setups				
Lysosensor yellow/blue (DND-160 or dextran)		~4.2	3–9	Yellow: 385/521 nm Blue: 335/452 nm
Lysosensor yellow/blue 10,000 MW dextran		~3.9	3–9	Yellow: 385/521 nm Blue: 335/452 nm
FITC dextran Texas-Red dextran		~6.4	5–8	FITC: 490/550 nm (pH sensitive) FITC: 440/550 nm TR: 596/615 nm
FITC dextran	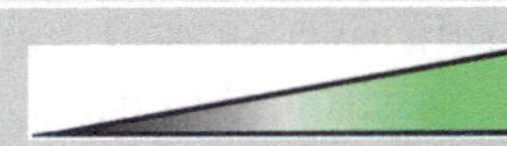	~6.4	5–8	490/550 nm (pH sensitive) 440/550 nm
Oregon green 488	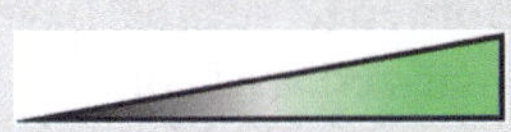	~4.7	4.2–5.7	490/520 nm (pH sensitive) 440/520 nm
Oregon green 514		~4.7		510/530 nm (pH sensitive) 450/530 nm
pHrodo-dextran red		N.A[a]	2–9	Red: 560/585 nm
pHrodo-dextran green		N.A[a]	2–9	Green: 509/533 nm

[a]pHrodo exhibits a fluorescence increase in response to decreasing pH, but in a nonlinear fashion, with the pH response profile changing upon what biomolecule the dye is conjugated to

methylamine, dihydrochloride (DAMP), and the frequently used LysoTracker and LysoSensor DND probes (Table 1) [31, 32]. There are several caveats to the use of acidotropic probes; first, they are nonselective and can accumulate in acidic compartments that are not lysosomes. Second, due to their basic nature, these probes can induce an alkalinizing effect on the lysosome. Lastly, acidotropic dyes are an indirect estimation of pH, and the localization can be easily affected by changes in membrane permeability, endocytic trafficking, cell size, and the number and size of acidic vesicles [33–35]. This is especially relevant when comparing disease models to healthy controls. For example, many lysosomal storage disease models show an increase in LysoTracker due to disease-associated lysosomal swelling, but this does not necessarily indicate a change in pH [36, 37].

The ratiometric dual-fluorescence imaging approach relies on biomolecules that have both a pH-sensitive and pH-independent emission spectrum. Dual fluorescence can be achieved either by using a combination of two probes that have different pH sensitivities (e.g., Texas red dextran and FITC dextran), or a single molecule that has two different emission spectra—one of which is proton-independent (e.g., LysoSensor Yellow/Blue). This enables a ratiometric calculation that controls for experimental variables such as changes in probe concentration (which can easily be affected by population cell membranes), the rate of trafficking, and lysosomal volume. It also allows for changes in instrumentation, such as sensitivity and focal planes. Ratiometric dyes are often conjugated to biomolecules that are endocytosed and reliably trafficked via the endocytic pathway to accumulate in lysosomes, but do not themselves alter the pH. This approach was first reported by Ohkuma and Poole, using fluorescein isothiocyanate (FITC)-labeled dextran [9]. FITC-dextran fluorescence emission is pH-dependent (pH 5–8) when excited at 490 nm, but is significantly less pH-sensitive at 440 nm. Absolute pH values can be derived from an in situ calibration, whereby luminal pH of organelles is comparable to that of the extracellular buffer of a known pH. Other ratiometric setups have since been utilized, some of which are summarized in Table 1.

The ideal probe would be solely present in the lysosome, not induce lysosomal damage or pH changes, have an appropriate loading control, have a pKa in the lysosomal pH range (4.5–5), and respond rapidly to changes in lysosomal pH. Although ratiometric dyes are considered the "gold standard" method of quantifying lysosomal pH, there are still caveats to all current methods of measuring lysosomal pH, due to the heterogeneous nature of lysosomes in the cell. Lysosomal positioning and motility play roles in cellular homeostasis, with peripheral lysosomes indicated to have a more alkaline pH and a different composition, potentially in part through altered connectivity with biological pathways (e.g., altered lysosomal/endosomal fusion) [38]. This means that cells with altered lysosomal motility may have differential loading of probes that is not necessarily due to pH changes. It remains clear that critically assessing images alongside the generated data is key to understanding lysosomal pH changes.

Here, we demonstrate two different detection strategies with a ratiometric FITC/Texas Red dextran setup to detect an increased lysosomal pH in a familial Alzheimer's disease presenilin 1 (PSEN1) cell model. The 10,000 MW dextran-conjugates are actively endocytosed by fluid-phase endocytosis and get "trapped" in the lysosome due to the large size. This is followed by an appropriately timed chase to ensure the dextran probe is only present in the lysosome. The first strategy uses traditional fluorescence light microscopy alongside an in situ calibration curve to determine absolute pH measurements. Second, we used a high content imaging system to generate relative pH measurements alongside a bafilomycin control curve for calibration. Agreeing with previous publications [39–41], we report an increase in lysosomal pH in PSEN mutant cells. Fluorescent microscopy assessment showed PSEN1 A246E HFs to have a lysosomal pH of 4.9 compared to pH 4.1 in age-matched controls (Fig. 1a–c), and high content imaging shows a significant increase in the FITC/Texas Red

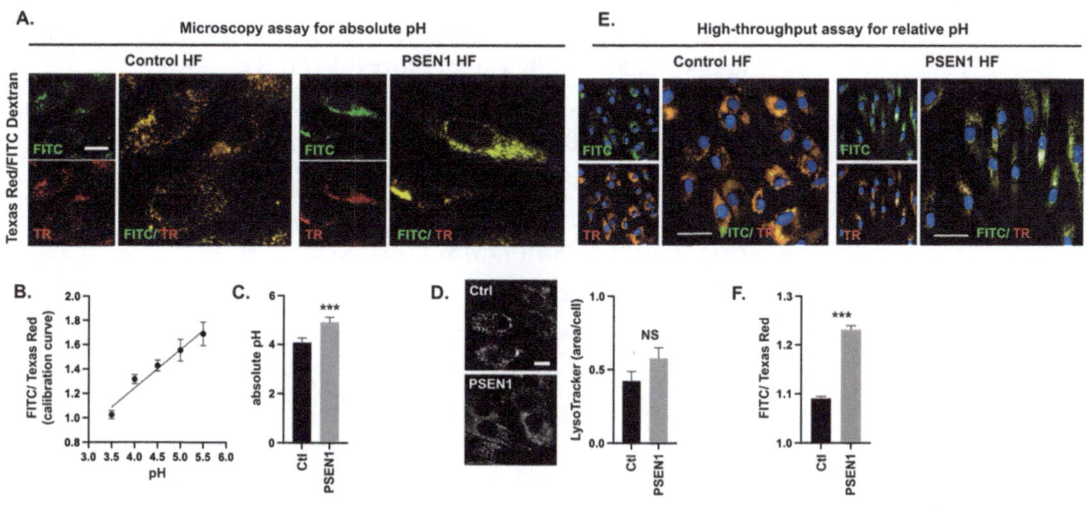

Fig. 1 Dual-imaging microscopy methods to demonstrate an increased lysosomal pH in PSEN1 patient fibroblasts compared to apparently healthy controls. (**a**) Representative images showing both control and PSEN1 human fibroblasts (HF) loaded with FITC-dextran (FITC) and Texas Red-dextran (TR) by a 24-h pulse, 24-h chase protocol to result in lysosomal localization. (**b**) A calibration curve generated from fluorescence intensity ratiometric measurements in control human fibroblasts, equilibrated with buffers at pH 3.5, 4.0, 4.5, 5.0, and 5.5. (**c**) Absolute pH values calculated by fitting the average fluorescent intensity measurements to the calibration curve. PSEN1 cells have an increased pH (4.90 $\pm$ 0.1) compared to control cells (4.07 $\pm$ 0.1). (**d**) Representative images of LysoTracker green in control and PSEN1 cells. Graphed data shows no significant difference in LysoTracker area between cell types, demonstrating the unsuitability of LysoTracker as a pH marker. (**e**) Representative images showing both control and PSEN1 human fibroblasts acquired on the Operetta high-content imaging system. (**f**) Ratiometric measurement of FITC/Texas Red Dextran as calculated via Operetta analysis. Representative scale bars = 10 μm; graphed data are presented as average $\pm$ SEM

dextran ratio, indicative of raised pH (Fig. 1e, f). Although PSEN1 lysosomes are less acidic, we also show no change in LysoTracker staining, highlighting the unsuitability of LysoTracker for pH measurements, and the necessity for loading controls (Fig. 1d).

2 Materials

2.1 Cell Lines and Maintenance

1. Apparently healthy human fibroblasts (Coriell Cell Repository, GM05399).
2. PSEN1 human fibroblasts carrying an A246E mutation (Coriell Cell Repository, AG06848).
3. DMEM complete cell growth medium: DMEM supplemented with 10% fetal bovine serum (PAN biotech), and 2 mM L-Glutamine.
4. Trypsin for cell detachment and passaging.
5. 8-well μ-slides (Ibidi, *see* **Note 1**).
6. 96-well Ultra Microplate (PerkinElmer, 6055302, *see* **Note 1**).

2.2 Buffers and pH Probes

1. Dextran, Texas Red™, 10,000 MW, Lysine Fixable (Invitrogen, D1863): 4× stock solution prepared in complete DMEM and stored at −20 °C (*see* **Notes 2** and **3**).
2. Fluorescein isothiocyanate-dextran 10,000 MW (Sigma-Aldrich, FD10S): 4× stock solution prepared in complete DMEM and stored at −20 °C (*see* **Notes 2** and **3**).
3. Bafilomycin A1 (Glentham Life Sciences, GA6363): stock solution of 200 mM in DMSO, stored at −20 °C.
4. Hoechst 33342 nucleic acid stain.
5. Complete HBSS: HBSS supplemented with 1 mM HEPES pH 7.4, 1 mM $MgCl_2$, and 1 mM $CaCl_2$.
6. Calibration buffers: 50 mM CH_3COONa, 5 mM NaCl, 115 nM KCl, 1.2 mM $MgSO_4$, 10 μM nigericin, 10 μM monesin (nigericin and monesin added directly prior to experiment). Individual calibration buffers are adjusted to pH 3.5, 4.0, 4.5, 5.0, and 5.5, using NaOH or HCl (*see* **Note 4**).
7. DMEM complete imaging medium: FluoroBrite™ DMEM supplemented with 10% fetal bovine serum, and 2 mM L-Glutamine.

2.3 Traditional Fluorescence Microscopy Imaging (See Note 5)

1. Microscope: Inverted widefield fluorescence or confocal with a minimum 40× oil or water immersion objective. This is needed to enable live cell imaging and to ensure that individual lysosomes can be resolved. An inverted microscope also enables

additions to be made over the course of an experiment and resultant changes in lysosomal pH to be followed in real time.

2. Light source: LED, HXP lamp, or lasers with appropriate excitation wavelengths or filters for optimal excitation of the probe, coupled with appropriate emission filters to allow detection of probe fluorescence.
3. Camera: sCMOS or CCD camera with sufficient sensitivity to detect changes in probe fluorescence.
4. Software: Imaging software compatible with the microscope, illumination system, and camera used that can record grayscale images, identify regions of interest, and quantify gray values within these regions. Suitable programs include Zen (Zeiss), LAS X (Leica), MetaFluor (Molecular Devices), Volocity (Quantum Technologies), and Micromanager (NIH).

2.4 High-Content Imaging Microscopy Imaging

1. Operetta CLS high-content imaging system (Perkin Elmer): equipped with Harmony software (Revvity) and appropriate LED illumination and emission filters (*see* **Note 6**).

3 Methods

3.1 FITC-Dextran/ Texas Red Dextran Light Microscopy Assay (See Note 5)

For light microscopy pH quantification, prior to dye loading, fibroblast cells were grown in 8-well μ-slides containing 250 μL complete DMEM. Cells were left overnight, or until a confluency of approximately 80% was achieved.

3.1.1 Loading Dextran Probes

1. 48 h prior to pH measurements (*see* **Note** 7), load cells with 250 μg/mL Texas Red dextran and 250 μg/mL FITC-dextran, in 250 μL complete DMEM (*see* **Notes 3** and **8**).
2. 24 h post addition of dextran (*see* **Note** 7), remove the DMEM and wash the cells three times in 1× PBS.
3. Incubate cells in complete DMEM for a further 24 h.
4. Wash cells 3× in complete HBSS and leave in the third wash of 150 μL of complete HBSS for imaging.

3.1.2 Imaging and Analysis of Dextran Probes

1. Place the 8-well chamber slide on the inverted microscope stage and remove the chamber slide lid.
2. Using a 40× objective, focus on a field containing cells and ensure both FITC and Texas Red dextran are visible—loading should look punctate and colocalized (*see* **Note** 7).
3. Adjust the acquisition parameters of the software as required (binning, gain, exposure time, automated acquisition). Minimize bleed through between channels (e.g., use separate tracks on a confocal, if available).

4. Acquire fluorescence images (≅4–8) in the appropriate channels (Table 1) per condition. Exposure and other parameters may be adjusted as appropriate between experiments, but not between channels in the same frame.
5. After image acquisition, calculate the background-subtracted FITC/Texas Red fluorescent intensities and generate a ratio per image, and then generate an average per condition.
6. Fit the fluorescent intensity ratio to the calibration curve (*see* Subheading 3.1.3) to generate an absolute pH value.

3.1.3 pH Calibration, In Situ

Calibration is carried out to align fluorescence values to absolute pH values and must be performed for each separate experimental repeat. This method is not adapted for high-content imaging (*see* Subheading 3.2).

1. Load cell with FITC/Texas Red dextran as described above (*see* Subheading 3.1.1).
2. On the day of the experiment, prepare individual pH calibration buffers at pH 3.5, 4.0, 4.5, 5.0, and 5.5, and filter sterilize (*see* **Note 9**).
3. Remove the media from the dextran incubated cells and wash with calibration buffer of a known pH (e.g., pH 4.0).
4. Add nigericin and monensin to the calibration buffer of known pH to a final concentration of 10 μM of each.
5. Remove the calibration buffer from the cell and incubate cells with 200 μL calibration buffer (now containing the nigericin and monesin) for 10 min to alter the lysosomal pH to that of the calibration buffer.
6. The 10-min incubation should be sufficient for the lysosome to equilibrate to the external pH (*see* **Note 10**). Acquire images at the appropriate ex/em parameters (e.g., FITC ex/em: 490/520, TR ex/em: 595/615), taking 5 images quickly per well. When acquiring images, do not shift the focal plane between channels.
7. Repeat **steps 3–6** at each pH buffer to generate a full calibration curve (*see* **Note 9**).
8. Select the regions of interest from your images and generate a ratiometric fluorescence intensity value. Plot the average ratiometric-intensity value against the pH of the calibration buffer to generate a calibration curve.

3.2 FITC-Dextran/Texas Red Dextran High-Throughput Assay (96-Well Plate)

The purpose of a high-throughput Operetta plate reader assay is to determine changes in pH (e.g., in response to a particular compound) that can be used, for example, as part of a drug discovery cascade. This assay does not allow for pH quantification but instead for a comparative measure, as the time taken to image the whole plate is too long for the cells to survive in an acidic pH calibration buffer. In lieu of a calibration curve, a bafilomycin control curve is generated to compare between experiments.

3.2.1 Assay Day 1

1. Trypsinize a confluent T75 flask of cells at 37 °C, for no longer than 5 min, until the cells are no longer adhering to the flask.
2. While the cells are trypsinizing, dilute the 4× stock solution with complete DMEM to prepare a DMEM-dextran solution containing 200 μg/mL FITC-dextran and 100 μg/mL Texas Red-dextran (100 μL required per well of a 96-well plate; *see* **Note 11**).
3. Once cells have been sufficiently trypsinized, add 6 mL growth media, and disperse the cells by pipetting.
4. Centrifuge for 5 min at 0.2× *g*. Discard the supernatant and resuspend the cells in 1 mL complete DMEM.
5. Count live cells and dilute with DMEM–dextran to get a seeding solution of 4×10^4 cells/mL. Add 100 μL of this cell suspension to each well.
6. Each plate requires 18 wells to generate a Bafilomycin control curve (1500, 750, 375, 187.5, 93.75, 46.88, 23.44, 11.72, 5.86, 2.93 nM; *see* **Note 12**). Add the Bafilomycin first at the highest concentration and perform a twofold dilution series across the wells.
7. If adding test compounds, add your drug and associated control to the plate at the appropriate times (*see* **Note 13**).
8. Leave plates for 24 h at 37 °C in 5% CO_2 after adding the DMEM–dextran (*see* **Note 7**).

3.2.2 Assay Day 2

1. Aspirate the media and wash cells with 100 μL PBS per well. Replace with 100 μL fresh complete DMEM (no dextrans).
2. Add experimental compounds back to the plate at the desired concentration (e.g., Bafilomycin, DMSO).
3. Leave plates for 24 h at 37 °C in 5% CO_2 after adding the DMEM–dextran (*see* **Note 7**).

3.2.3 Assay Day 3

1. Take 10 mL of complete HBSS and add 4 μL Hoechst (keep hidden from light until needed).
2. Aspirate media from the 96-well plate and wash 3× with 100 μL complete HBSS/well.
3. Add 100 μL HBSS with Hoechst and incubate at RT for 20 min.
4. Wash the plate 3× with HBSS, and transfer to complete FluorоBrite DMEM for Operetta imaging (*see* **Note 14**).

3.2.4 Operetta Imaging and Analysis

1. Set up environmental control for the equipment (5% CO_2, 37 °C).
2. Run a nuclear pre-scan with the Hoechst filter setting to identify fields of view that contain cells.
3. Image cells using the Hoechst, FITC, and Texas-Red built-in filter settings in non-confocal mode using the 63× water immersion objective. Check several wells before choosing the best plane of focus, LED intensity, and exposure time for each wavelength. Capture at least 10 images per well.
4. To analyze, first use the Hoechst channel to identify cells and define the area of the cell (*see* **Note 15**).
5. Use the Texas Red channel to find spots within the identified "cells" population.
6. Measure FITC and Texas Red intensity within the identified spots.
7. Calculate a ratio for FITC/Texas Red by object—this generates a ratiometric spot intensity measure for every spot in a cell. This will allow comparison of the average FITC/Texas Red ratio between conditions.
8. Check the Bafilomycin control curve to assess for experimental consistency. Plot a concentration response curve and check the EC_{50} value to be within twofold of the running average across experiments. Calculate the Z' between the positive (highest bafilomycin concentration) and negative (DMSO only) controls to get a value of at least 0.5 (generally at least a twofold difference in the ratio).

4 Notes

1. Imaging plates may be optimized for your setup but must have a suitable material on the base for optical imaging.
2. Dextran-conjugates offer several advantages over membrane-permeant probes; first, they are reliably trafficked to lysosomes via the endocytic pathway, while membrane-permeant probes can label other acidic compartments that are not lysosomes (e.g., late endosomes). Second, dextran conjugates do not appear to induce osmotic swelling and can be used at a relatively low concentration due to the high sensitivity that comes from multiple fluorophores conjugated to a single dextran probe.
3. The exact concentration of any lysosomal probe must be optimized per cell type. The concentrations given in the text are for the cell types used here and may need to be dramatically altered for use in other biological models. The optimal quantity is the

lowest concentration at which there is still a reliable signal. Too high can cause saturation, nonspecific loading, and effects on lysosomal biology, such as alkalinization.

4. A new calibration curve must be carried out when new stock solutions of lysosomal probe are made. Stock solutions must be treated equally to those used to make the calibration curve (e.g., same number of freeze-thaw cycles).
5. Assessment by microscopy has several advantages over high-throughput imaging. First, microscopy allows for probe calibration, so that pH can be accurately quantified. Because of the time needed to read a whole high-content imaging plate, this is much more difficult to achieve using a high-throughput imaging approach. Furthermore, traditional microscopy allows for focal adjustment for each field of view to ensure the lysosomes are always in focus, and for changes in lysosomal pH following the addition of compounds to be followed in real time.
6. High throughput analysis of lysosomal pH by high content imaging can utilize any suitable system, e.g., Opera Phenix or Operetta (Perkin Elmer), Cell Discoverer (Zeiss), ImageXpress (Molecular Devices), Cellsnight (Thermo Scientific)—providing it has illumination (laser, LED) and emission filters at appropriate wavelengths for the probes can be used, and associated analysis software. These systems should be equipped with temperature and CO_2 control to avoid changes in pH of the culture medium caused by loss of CO_2.
7. The pulse/chase time for dextran needs to be optimized for individual cell types, as it is highly influenced by the speed of endocytosis. The pulse needs to ensure the probe has had enough time to reach the lysosome, and the chase needs to make sure the dextran has had time to accumulate in the terminal lysosome—as opposed to being retained in the endosomal trafficking system or recycled out through recycling endosomes. This is especially important when comparing cells; for example, disease can greatly affect the rate of endocytosis. Endocytic probes can be checked for lysosomal loading by using a fixable dextran conjugate and co-staining with lysosomal marker LAMP1. Lysosomal loading could also be checked via co-staining with LysoTracker—although this is less accurate, as we know LysoTracker can stain some non-lysosomal compartments.
8. For dual emission, FITC-dextran can be coloaded with pH-insensitive Texas red dextran. A pitfall of using two fluorescent signals that do not originate from the same fluorophore is the potential for differential photobleaching of the two fluorophores, and differential trafficking speed resulting in differential loading—although this is mitigated by using the same size of conjugated dextran.

9. The calibration curve needs to be broad enough to cover all the measured values within an experiment—you cannot extrapolate as the relationship of the probes with pH is not always linear.
10. When the cells have reached equilibrium with the calibration buffer, there should be no large fluorescence fluctuations occurring. To check this, focus on a test region and take a photo once every 30 s for 2 min. If there are dramatic changes in fluorescence, test longer and shorter incubation times with the calibration buffer.
11. The concentration of lysosomal probe must also be optimized for the imaging modality; for example, we utilize a higher concentration of FITC dextran in the Operetta high-content screening, as opposed to the inverted microscope setup.
12. Bafilomycin can, by inhibiting vATPase [42], be used to mimic disease phenotypes that deacidify the lysosome. Adding a bafilomycin-induced deacidification curve to each plate allows for plate-to-plate comparison—the concentrations of bafilomycin should be optimized per cell type. Artesunate can be used as an acidification control compound as it increases vATPase activity—although this compound is toxic at high concentration, meaning a full IC_{50} curve cannot be generated.
13. To assess lysosomal re-acidification, cells can be pre-treated or co-treated with a low concentration (approx. 50 nM) of Bafilomycin, and the results compared back to the Bafilomycin control curve.
14. A temperature and CO_2-controlled microscope environment is preferred to minimize variability, although it is not essential, as HBSS can maintain physiological pH in a non-CO_2 atmosphere. If CO_2 is used, an alternative imaging medium may be preferable. We have observed more consistent results and brighter signals when using FluoroBrite™ DMEM during environmentally controlled automated imaging.
15. Generally, Hoechst provides sufficient low-intensity background staining to allow the Operetta to assign cell outlines. If this is not effective in your automated setup, brightfield or digital phase contrast imaging can be used to define cell areas and exclude non-intracellular background fluorescence from the analysis.

Acknowledgments

EDK was supported by an Alzheimer's Research UK collaborator grant (ARUK-IRG2015-7) awarded to ELE. SEG was supported by a BRACE Dementia Research PhD studentship (BR2240) awarded to HWE.

References

1. Ballabio A, Bonifacino JS (2020) Lysosomes as dynamic regulators of cell and organismal homeostasis. Nat Rev Mol Cell Biol 21(2): 101–118
2. Saftig P, Klumperman J (2009) Lysosome biogenesis and lysosomal membrane proteins: trafficking meets function. Nat Rev Mol Cell Biol 10(9):623–635
3. De Duve C et al (1955) Tissue fractionation studies. 6. Intracellular distribution patterns of enzymes in rat-liver tissue. Biochem J 60(4): 604–617
4. Colacurcio DJ, Nixon RA (2016) Disorders of lysosomal acidification-the emerging role of v-ATPase in aging and neurodegenerative disease. Ageing Res Rev 32:75–88
5. Lloyd-Evans E, Haslett LJ (2016) The lysosomal storage disease continuum with ageing-related neurodegenerative disease. Ageing Res Rev 32:104–121
6. Mindell JA (2012) Lysosomal acidification mechanisms. Annu Rev Physiol 74:69–86
7. Harikumar P, Reeves JP (1983) The lysosomal proton pump is electrogenic. J Biol Chem 258(17):10403–10410
8. Dell'Antone P (1979) Evidence for an ATP-driven "proton pump" in rat liver lysosomes by basic dyes uptake. Biochem Biophys Res Commun 86(1):180–189
9. Ohkuma S, Moriyama Y, Takano T (1982) Identification and characterization of a proton pump on lysosomes by fluorescein-isothiocyanate-dextran fluorescence. Proc Natl Acad Sci USA 79(9):2758–2762
10. Van Dyke RW (1993) Acidification of rat liver lysosomes: quantitation and comparison with endosomes. Am J Phys 265(4 Pt 1):C901–C917
11. Chakraborty K, Leung K, Krishnan Y (2017) High lumenal chloride in the lysosome is critical for lysosome function. elife 6
12. Deriy LV et al (2009) Disease-causing mutations in the cystic fibrosis transmembrane conductance regulator determine the functional responses of alveolar macrophages. J Biol Chem 284(51):35926–35938
13. Di A et al (2006) CFTR regulates phagosome acidification in macrophages and alters bactericidal activity. Nat Cell Biol 8(9):933–944
14. Graves AR et al (2008) The Cl-/H+ antiporter ClC-7 is the primary chloride permeation pathway in lysosomes. Nature 453(7196):788–792
15. Wu JZ et al (2023) ClC-7 drives intraphagosomal chloride accumulation to support hydrolase activity and phagosome resolution. J Cell Biol 222(6)
16. Zhang Q et al (2023) Lysosomal chloride transporter CLH-6 protects lysosome membrane integrity via cathepsin activation. J Cell Biol 222(6)
17. Steinberg BE et al (2010) A cation counterflux supports lysosomal acidification. J Cell Biol 189(7):1171–1186
18. Lee D, Hong JH (2023) Modulation of lysosomal Cl(−) mediates migration and apoptosis through the TRPML1 as a lysosomal Cl(−) sensor. Cells 12(14)
19. Moriyama Y (1988) Potassium ion dependent proton efflux and depolarization from spleen lysosomes. Biochem Biophys Res Commun 156(1):211–216
20. Jinn S et al (2019) Functionalization of the TMEM175 p.M393T variant as a risk factor for Parkinson disease. Hum Mol Genet 28(19):3244–3254
21. Zheng W et al (2022) pH regulates potassium conductance and drives a constitutive proton current in human TMEM175. Sci Adv 8(12): eabm1568
22. Zhang J et al (2023) Lysosomal LAMP proteins regulate lysosomal pH by direct inhibition of the TMEM175 channel. Mol Cell 83(14): 2524–2539 e7
23. Fujii T et al (2023) Parkinson's disease-associated ATP13A2/PARK9 functions as a lysosomal H,K-ATPase. Nat Commun 14(1)
24. Gharanei S et al (2013) Vacuolar-type H+-ATPase V1A subunit is a molecular partner of Wolfram syndrome 1 (WFS1) protein, which regulates its expression and stability. Hum Mol Genet 22(2):203–217
25. Williamson WR, Hiesinger PR (2010) On the role of v-ATPase V0a1-dependent degradation in Alzheimer disease. Commun Integr Biol 3(6):604–607
26. Dehay B et al (2013) Lysosomal impairment in Parkinson's disease. Mov Disord 28(6): 725–732
27. Gotzl JK et al (2014) Common pathobiochemical hallmarks of progranulin-associated frontotemporal lobar degeneration and neuronal ceroid lipofuscinosis. Acta Neuropathol 127(6):845–860
28. Jiang Y et al (2019) Lysosomal dysfunction in down syndrome is APP-dependent and mediated by APP-betaCTF (C99). J Neurosci 39(27):5255–5268

29. de Duve C et al (1974) Commentary. Lysosomotropic agents. Biochem Pharmacol 23(18): 2495–2531
30. MacIntyre AC, Cutler DJ (1988) The potential role of lysosomes in tissue distribution of weak bases. Biopharm Drug Dispos 9(6):513–526
31. Allison AC, Young MR (1962) Uptake of dyes and drugs by living cells in culture. Life Sci 1964(3):1407–1414
32. Chazotte, B., Labeling lysosomes in live cells with LysoTracker. Cold Spring Harb Protoc, 2011. 2011(2): pdb prot5571
33. Palmgren MG (1991) Acridine orange as a probe for measuring pH gradients across membranes: mechanism and limitations. Anal Biochem 192(2):316–321
34. Tsien RY (1989) Fluorescent indicators of ion concentrations. Methods Cell Biol 30:127–156
35. Pierzynska-Mach A, Janowski PA, Dobrucki JW (2014) Evaluation of acridine orange, LysoTracker Red, and quinacrine as fluorescent probes for long-term tracking of acidic vesicles. Cytometry A 85(8):729–737
36. Lloyd-Evans E et al (2008) Niemann-Pick disease type C1 is a sphingosine storage disease that causes deregulation of lysosomal calcium. Nat Med 14(11):1247–1255
37. te Vruchte D et al (2014) Relative acidic compartment volume as a lysosomal storage disorder-associated biomarker. J Clin Invest 124(3):1320–1328
38. Johnson DE et al (2016) The position of lysosomes within the cell determines their luminal pH. J Cell Biol 212(6):677–692
39. Lee JH et al (2010) Lysosomal proteolysis and autophagy require presenilin 1 and are disrupted by Alzheimer-related PS1 mutations. Cell 141(7):1146–1158
40. Wolfe DM et al (2013) Autophagy failure in Alzheimer's disease and the role of defective lysosomal acidification. Eur J Neurosci 37(12):1949–1961
41. Coffey EE et al (2014) Lysosomal alkalization and dysfunction in human fibroblasts with the Alzheimer's disease-linked presenilin 1 A246E mutation can be reversed with cAMP. Neuroscience 263:111–124
42. Lee JH et al (2015) Presenilin 1 maintains lysosomal Ca(2+) homeostasis via TRPML1 by regulating vATPase-mediated lysosome acidification. Cell Rep 12(9):1430–1444

Chapter 3

Assessing Perturbations in Lysosome Health by Flow Cytometry in iPSC-Derived Human Neuron Cultures

Kirstin O. McDonald, Stephanie M. Hughes, and Indranil Basak

Abstract

Fluorescent molecular probes have frequently been used to monitor lysosomal health, localization, abundance, and movement through the detection of acidic organelles and lysosomal enzyme activity. Flow cytometry technology provides rapid and accurate analysis of single cells (neurons) or particles (lysosomes) in suspension through laser detection. Herein, we describe how to detect lysosomes via LysoTracker™ and Magic Red® Cathepsin Activity assays in iPSC-derived human neuron cultures by flow cytometry.

Key words Lysosome function, Lysosome acidity, Fluorescent probes, Flow cytometry, iPSC-derived neuron cultures

1 Introduction

Lysosomes are made up of a series of acid hydrolases encapsulated in a single carbohydrate-rich phospholipid bilayer. An acidic milieu is required for the acid hydrolases to be able to perform their function [1–3]. Lysosomes are particularly important for neurons, given their unique shape and polarity. As lysosomes move from the soma to the axon, both acidity and hydrolytic activity decrease [4–7]. Disruptions to lysosome health are cataclysmic and result in lysosomal storage disorders (LSDs), metabolic disorders, neurodegenerative disorders, and cancer. The incorporation and accumulation of nontoxic fluorescent dyes in the acidic environment of the lysosome offer an efficient way to detect changes in lysosomal acidity, function, localization, movement, and trafficking [8–13].

LysoTracker is a fluorescent probe that accumulates in acidic compartments within a cell, such as lysosomes. The LysoTracker probe is a weak hydrophobic base with several aromatic rings that can freely diffuse across cell membranes [14]. When the Lysotracker probe encounters the acidic environment of the lysosome, it becomes protonated. In this charged state, the probe is no longer

Stephanie M. Hughes and Indranil Basak (eds.), *Lysosomes: Methods and Protocols*, Methods in Molecular Biology, vol. 2976, https://doi.org/10.1007/978-1-0716-4844-5_3,

able to cross the organelle membrane, providing a localized pH-dependent accumulation for distinct staining of acidic organelles.

Magic Red® is a cell-permeable substrate for the active cathepsin enzymes in the lysosome. Detection substrates utilize the photostable red fluorophore, cresyl violet. When quenched, the cresyl violet is nonfluorescent. Following enzymatic cleavage by Cathepsin B at one or both arginine (R) amide linkage sites, the mono and nonsubstituted cresyl violet fluorophores generate red fluorescence.

Coupling of lysosome-localized and acid-responsive dyes has led to the ability to screen for changes in lysosome acidity by flow cytometry [15]. This presents a fully unbiased quantitative method to detect changes in lysosome acidity, function, abundance, and size to complement commonly used microscopy-based techniques [16, 17]. In addition, it also provides opportunities to perform CRISPR-based screens for genes and drug-based screens that alter lysosome acidity and function [18, 16, 19]. Alternative techniques, such as LysoSensor or FITC, which utilize dextran to enter the cell, can be used to quantify lysosomal pH [3, 20–23]. However, the lysotracker and magic red assays detected via flow cytometry offer a high-throughput technique to efficiently assess changes in lysosome health. Herein, we detail a protocol for assessing changes in lysosome acidity and enzyme function by flow cytometry using the fluorescent probes LysoTracker™ Red DND-99 and Magic Red® Fluorescent Cathepsin B in induced pluripotent stem cell (iPSC) -derived human glutamatergic cortical neurons.

2 Materials

Prepare all solutions using cell culture-grade reagents and sterile cell culture techniques. Where possible, all buffers should be prepared in a laminar flow/cell culture hood. Buffers should be stored at room temperature unless otherwise indicated.

2.1 Cell Culture Plating, and Maintenance

1. Borate buffer: Boric acid 100 mM, sodium tetraborate 25 mM, sodium chloride 75 mM, pH 8.4. Add 100 mL of cell culture-grade water to an appropriately sized measuring cylinder. Weigh out 3.09 g of boric acid, 4.77 g of sodium tetraborate, and 2.19 g of sodium chloride and transfer to a cylinder. Add water to 450 mL and mix components well. Adjust pH with sodium hydroxide as needed to reach the desired pH of 8.4. Top up water to 500 mL.
2. Poly-L-ornithine (PLO) 100 μg/mL: To an unopened bottle of PLO (50 mg), add 5 mL of filtered borate buffer to generate a 10 mg/mL stock. Store at −20 °C in aliquots. To prepare a

working solution, dilute 1 in 100 in water, i.e., 10 μL of poly-L-ornithine for every 1 mL of water. Filter sterilize before use.

3. Dulbecco's phosphate buffered saline (dPBS): dPBS is modified without calcium and magnesium and is at pH 7.4. Filter sterilize before use.
4. Cell culture grade distilled water: Filter sterilize before use.
5. Doxycycline 2 μg/mL: Add 20 mg of powdered doxycycline to 10 mL of filtered PBS to generate a 2 mg/mL stock. Store at −20 °C in aliquots (*see* **Note 1**). To prepare a working solution, dilute 1 in 1000 in cell culture medium, i.e., 1 μL for every 1 mL of culture media for a 2 μg/mL working solution.
6. Brain Derived Neurotrophic Factor (BDNF): 10 ng/mL BDNF in PBS supplemented with 0.1% IgG and protease-free BSA. Prepare a solution of 0.1% IgG and protease-free BSA by weighing out 5 mg of IgG and protease-free BSA. Add 5 mL of PBS and mix to dissolve and then filter sterilize. To an unopened bottle of BDNF (50 μg), add 5 mL of the freshly prepared PBS supplemented with 0.1% IgG and protease-free BSA to generate a 10 μg/mL stock. Store in aliquots at −80 °C. Prepare a 1:1000 working dilution by adding 1 μL of stock solution for every 1 mL of media.
7. Neurotrophin-3 (NT3): 10 ng/mL NT-3 in PBS supplemented with 0.1% IgG and protease-free BSA. Prepare PBS supplemented with 0.1% IgG and protease-free BSA as detailed above under BDNF. To an unopened bottle of NT-3 (50 mg), add 5 mL of freshly prepared PBS supplemented with 0.1% IgG and protease-free BSA to generate a 10 μg/mL stock. Store in 50 μL aliquots at −80 °C. Prepare a 1:1000 working dilution by adding 1 μL of stock solution for every 1 mL of media.
8. Laminin (*see* **Note 2**): Aliquot the commercial stock into 50 μL portions and store at −80 °C.
9. Cortical Neuron media: BrainPhys Neuronal media supplemented with B27, BDNF 10 ng/μL, NT-3 10 ng/μL, and laminin 1 μg/mL. Add 1 mL of B27, and 50 μL of each laminin, NT-3, and BDNF to 49 mL of BrainPhys. Filter sterilize and store at 4 °C. Preheat to 37 °C prior to adding to cells.
10. 24-well cell culture grade plates.
11. 37 °C (5% CO_2) cell culture incubator.
12. Trypan blue.

2.2 Cell Staining

1. LysoTracker™ Red: Aliquot 1 mM LysoTracker™ Red in anhydrous DMSO to generate smaller working aliquots (5–10 μL) and store at −20 °C (*see* **Note 3**).

2. Magic Red® Fluorescent Cathepsin B Assay Kit: Add 50 μL of DMSO to the lyophilized Magic Red and gently swirl to mix. The resulting stock solution will be 250× and should appear red. Aliquot into 10 tubes, each with 5 μL aliquots, and store at −20 °C for up to 6 months. Avoid more than two freeze-thaws.
3. Propidium Iodide (PI) (*see* **Note 4**) 50 μg/mL in PBS: Measure out 5 mL of filtered PBS in a 10-mL Falcon tube. Weigh out 100 mg of PI and dissolve in 5 mL of PBS to generate a 20 mg/mL stock solution. Mix well and store at −20 °C in aliquots. To generate the working solution, dilute the stock solution 1 in 400, i.e., 2.5 μL stock PI for every 1 mL of media.
4. Cortical neuron phenol red-free media: see above.
5. 37 °C (5% CO_2) cell culture incubator.

2.3 Flow Cytometry

1. Accutase: Aliquot and store at −20 °C wrapped in tinfoil to protect from light.
2. Cortical neuron media: phenol red free (see above).
3. FACS tubes.

3 Methods

All concentrations and cell densities have been optimized for iPSC-derived human glutamatergic neurons maintained in cortical neuron media on PLO-coated plates as previously described [8, 24]. For a 24-well plate, Day 0 cortical neurons are plated at a density of 100,000–200,000 cells per well. Optimization is recommended for other neuronal cell cultures, maintenance media, and cell densities.

3.1 Cell Plating and Maintenance

1. Add ½ culture volume of PLO to the tissue culture dishes to be used for plating Day 0 cortical neurons. Gently tilt the plate to ensure full coverage.
2. Incubate dishes for at least 1 h at room temperature or 37 °C (*see* **Note 5**).
3. Aspirate coating solution, then wash dishes 3× with cell culture grade filtered sterile water.
4. Aspirate water and let dishes dry completely in the cell culture hood (typically requires 30 min–1 h) (*see* **Notes 6** and **7**).
5. Prepare cortical neuron medium + doxycycline (1:1000 dilution) and heat in a 37 °C water bath until warm to the touch.
6. Ensure rapid thawing of frozen cells, and add 1 mL of prewarmed cortical neuron media.
7. Centrifuge at 300× *g* for 5 min.

8. Discard the supernatant and resuspend in 1 mL of prewarmed cortical neuron media.
9. Dilute cells 1:1 with Trypan Blue, i.e., 10 μL of cell suspension with 10 μL of Trypan Blue.
10. Calculate the cell number.
11. Plate 100,000 cells per well in a total volume of 500 μL in prewarmed cortical neuron media.
12. Gently disperse cells for even coverage of the well by shaking.
13. Observe cells under a phase contrast microscope (*see* **Note 8**).
14. Perform bi-weekly half-media changes (i.e., every 3–4 days) until the day of the LysoTracker™ Red DND-99 assay or Magic Red® Fluorescent Cathepsin B assay day (*see* **Note 9**).

3.2 Cell Staining

All concentrations have been optimized for iPSC-derived human glutamatergic cortical neurons. The LysoTracker™ Red DND-99 assay and Magic Red® Fluorescent Cathepsin B assay are performed in cortical neuron maintenance media. Resuspension in phenol red-free cortical neuron media for flow cytometry analysis is recommended.

3.2.1 Lysotracker™ Red Staining

This protocol enables monitoring of changes in acidic organelles via the LysoTracker™ Red DND-99 assay kit. Please note that the kit is also available with a green fluorophore. The controls recommended for the lysotracker assay include Bafilomycin, which inhibits the vATPase and autophagosome/lysosome fusion, and/or Chloroquine, which localizes to the lysosome and causes an increase in lysosomal pH.

1. Dilute LysoTracker™ Red to 1:100 in cell culture media, i.e., 1 μL in 99 μL of prewarmed cortical neuron media.
2. From the initial dilution, do a further 1:200 dilution to give a final dilution of 1:20,000 (50 nM final concentration), e.g., 10 μL in 1990 μL of cortical neuron media.
3. Discard the media from the cells and add 400 μL of the LysoTracker™ mix from **step 2** to the wells.
4. Incubate the cells in the cell culture incubator at 37 °C for 30 min.
5. Aspirate LysoTracker™ mix and gently wash with 500 μL of dPBS per well (*see* **Note 10**).
6. Aspirate dPBS and add 300 μL of Accutase. (NB. If also performing PI staining, switch to PI staining protocol (*see* Subheading 3.2.3) from here prior to lifting cells.)
7. Incubate at 37 °C for 5 min.
8. Centrifuge at 300× *g* for 5 min.

9. Aspirate supernatant and resuspend the cells in 400 μL of phenol red-free cortical neuron media and proceed for analysis (*see* **Notes 11** and **12**).

3.2.2 Magic Red® Staining

This protocol enables monitoring of intracellular cathepsin B activity via the Magic Red® Fluorescent Cathepsin B Assay Kit [10]. The cathepsin B selective inhibitor CA-074 is recommended as a control for the Magic Red® assay.

1. Prior to staining the samples, dilute the stock (250×) 1:10 in cortical neuron media to generate 25× 1° (primary) staining solution. Use the staining solution within 15 min to prevent substrate hydrolysis. A further 1:25 dilution in cortical neuron media is required to obtain the final working staining solution.
2. Discard old media and replace with 400 μL of the final 1× working staining solution from **step 2**.
3. Incubate the cells with the staining solution for 30 min at 37 ° C.
4. Aspirate Magic Red® stain and add up to ½ culture volume of Accutase. (NB. If also performing PI staining, switch to PI staining protocol (*see* Subheading 3.2.3) from here prior to lifting cells.) (*See* **Note 10**.)
5. Incubate at 37 °C for 5 min.
6. Centrifuge at 300× *g* for 5 min.
7. Aspirate supernatant and resuspend the cells in 400 μL of phenol red-free cortical neuron media and proceed for flow cytometry analysis (*see* **Note 11**).

3.2.3 Propidium Iodide Staining (Optional)

1. Prepare PI stocks of 20 mg/mL in PBS.
2. Dilute PI to a final concentration of 50 μg/mL, i.e., 1:400 dilution.
3. Add 500 μL of PI staining to each sample.
4. Incubate the cells in the dark for 20 min.
5. Aspirate PI stain and add up to ½ culture volume of Accutase and incubate at 37 °C for 5 min.
6. Centrifuge at 300× *g* for 5 min.
7. Collect cells in a FACS tube and proceed to flow cytometry analysis.

3.3 FACS Analysis

As per the manufacturer's details, LysoTracker™ Red can be detected with a TRITC filter. It is advised to perform optimization on your available flow cytometer to ensure LysoTracker™ Red is being detected optimally for your experiment. In this protocol, Lysotracker™ Red (Ex 577/590) and Magic Red® (Ex:

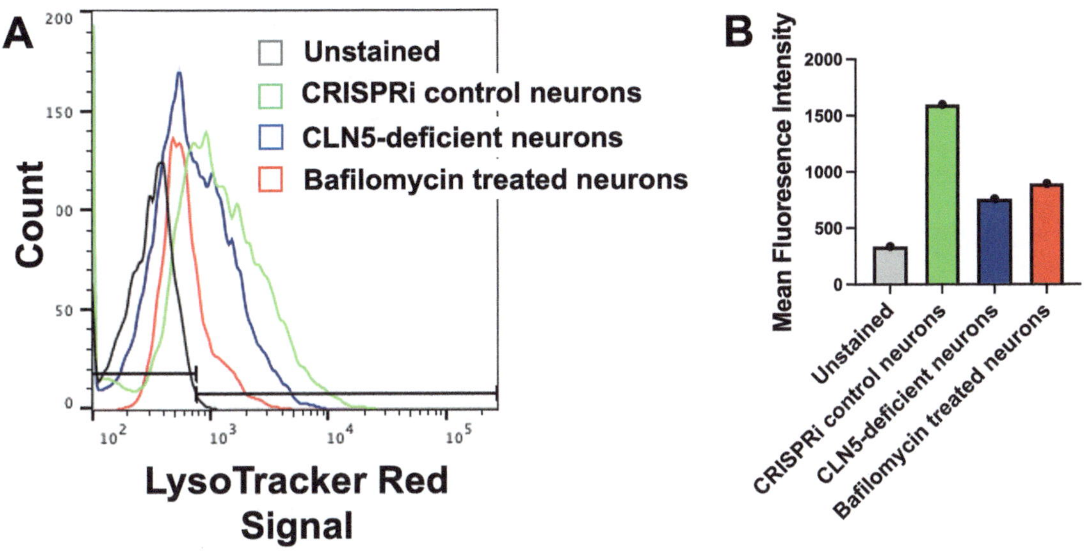

Fig. 1 Flow cytometric analysis of Lysotracker Red on iPSC-derived human neurons. (**a**) Flow cytometry analysis using FlowJo showing a shift in the histogram for LysoTracker Red signal. The black line is for unstained neurons, the green line is the control neurons showing maximum LysoTracker Red signal, followed by reduced LysoTracker Red signal for CLN5-deficient neurons (blue line) and bafilomycin-treated neurons (red line). (**b**) Measurement of Mean Fluorescence Intensity of the histograms quantifying LysoTracker signal in the neurons

550/590) are excited using a yellow-green (561 nm) laser with a YG-586/15 filter on a BD Fortessa Flow cytometer.

1. To observe the flow cytometric profile of the cells, plot a forward scatter (FSC, determining the size of the cells) versus a side scatter (SSC, determining the granularity of the cells). As an optional step, to confirm the live/dead cell population, a propidium iodide staining can be performed.
2. Generate a histogram count of YG-586/15+ (LysoTracker™ Red or Magic Red® positive) YG-586/15- (LysoTracker™ Red or Magic Red® negative) populations on the flow cytometer (Fig. 1A).
3. A total of 10,000 events are recorded for each sample.
4. Calculate the mean fluorescent intensity (MFI) for each YG-586/15+ sample and compare between treatment groups (Fig. 1B).

4 Notes

1. Doxycycline is unstable and should not undergo repeated freeze-thaw cycles. If using regularly, discard after 3–5 freeze-thaw cycles. Doxycycline drives the expression of neurogenin2

(mNgn2) to differentiate iPSCs into i3Ns. Doxycycline is used on Day 0 or plating day to ensure the differentiation of cells. However, prolonged Ngn2 expression hinders neuron maturation and therefore doxycycline must be diluted out through half media changes.

2. If aliquoting laminin, ensure tips and tubes have been prechilled.
3. Lysotracker is a light-sensitive dye and should be stored in tinfoil in a light-free box and used in a minimum light environment.
4. Propidium iodide is a hazardous chemical and should be weighed and resuspended in a safety cabinet or while wearing personal protective equipment, including a mask and safety goggles.
5. For best results, incubate dishes in PLO overnight in a 37 °C incubator.
6. To accelerate the drying process, stand the dishes on their sides and lean them against the back of the hood. Lids may also be left askew to allow better airflow.
7. Coated and dried dishes should be used immediately or wrapped in aluminum foil and stored at 4 °C for up to 1 week.
8. Check cells every day under a phase contrast microscope, paying particular attention to cell debris, morphological changes, and undifferentiated cells. High levels of cell debris and/or cell clumping often indicate a problem with either the dish coating or culture medium. If seen, remove half volume of culture medium and replace to full volume with fresh cortical neuron media (50% additional medium). If additional fresh medium does not result in less debris the next day, there has likely been insufficient coating or the coating medium was expired or the induction was not sufficient or there was a problem with the cortical neuron media.
9. Biweekly half-media changes (i.e., every 3–4 days) can be effective in sustaining dense cultures. After plating i3Ns on Day 0, do half media changes on Days 3, 7, 10, 17, and so on. Weekly half-media changes are sufficient to sustain long-term cultures at moderate densities. Use the appropriate serological pipette or micropipette to slowly aspirate a measured volume from each well, and very gently replace with fresh media. Neurons tend to easily dissociate from the dish, so any aspiration or dispensing of media directly onto cells is not recommended. Take care to aspirate and dissociate by tilting the dish so that the media accumulates on one side. Then, aspirate/dispense with the pipette directed toward the wall of the dish (i.e., away from the cells at the bottom).

10. Prior to lifting while the cells are in PBS, it is recommended to use a fluorescent microscope to ensure staining has been successful.
11. For normal maintenance and staining, BrainPhys phenol red media is used. For applications such as microscopy and flow cytometry analysis, BrainPhys phenol red-free media is used, as the incorporated fluorophore is also red.
12. LysoTracker™ Red staining only lasts for 3–4 h; therefore, proceed with Lysotracker detection immediately.

Acknowledgments

We thank the Neurological Foundation of New Zealand for supporting the development of the iPSC-derived CLN5 Batten disease model (1737 PG).

References

1. Saftig P, Schroder B, Blanz J (2010) Lysosomal membrane proteins: life between acid and neutral conditions. Biochem Soc Trans 38(6): 1420–1423. https://doi.org/10.1042/BST0381420
2. Coffey JW, De Duve C (1968) Digestive activity of lysosomes. I. The digestion of proteins by extracts of rat liver lysosomes. J Biol Chem 243(12):3255–3263
3. Ohkuma S, Poole B (1978) Fluorescence probe measurement of the intralysosomal pH in living cells and the perturbation of pH by various agents. Proc Natl Acad Sci USA 75(7): 3327–3331. https://doi.org/10.1073/pnas.75.7.3327
4. Ferguson SM (2019) Neuronal lysosomes. Neurosci Lett 697:1–9. https://doi.org/10.1016/j.neulet.2018.04.005
5. Gowrishankar S, Yuan P, Wu Y, Schrag M, Paradise S, Grutzendler J, De Camilli P, Ferguson SM (2015) Massive accumulation of luminal protease-deficient axonal lysosomes at Alzheimer's disease amyloid plaques. Proc Natl Acad Sci USA 112(28):E3699–E3708. https://doi.org/10.1073/pnas.1510329112
6. Hollenbeck PJ (1993) Products of endocytosis and autophagy are retrieved from axons by regulated retrograde organelle transport. J Cell Biol 121(2):305–315. https://doi.org/10.1083/jcb.121.2.305
7. Lee S, Sato Y, Nixon RA (2011) Lysosomal proteolysis inhibition selectively disrupts axonal transport of degradative organelles and causes an Alzheimer's-like axonal dystrophy. J Neurosci 31(21):7817–7830. https://doi.org/10.1523/JNEUROSCI.6412-10.2011
8. Basak I, Hansen RA, Ward ME, Hughes SM (2021) Deficiency of the lysosomal protein CLN5 alters lysosomal function and movement. Biomolecules 11(10). https://doi.org/10.3390/biom11101412
9. Prabhu AV, Kang I, De Pace R, Wassif CA, Fujiwara H, Kell P, Jiang X, Ory DS, Bonifacino JS, Ward ME, Porter FD (2021) A human iPSC-derived inducible neuronal model of Niemann-Pick disease, type C1. BMC Biol 19(1):218. https://doi.org/10.1186/s12915-021-01133-x
10. Magic Red® Fluorescent Cathepsin B Assay Kit. https://www.immunochemistry.com/products/magic-red-cathepsin-b-assay-kit-937
11. Pal A, Glass H, Naumann M, Kreiter N, Japtok J, Sczech R, Hermann A (2018) High content organelle trafficking enables disease state profiling as powerful tool for disease modelling. Sci Data 5:180241. https://doi.org/10.1038/sdata.2018.241
12. Gowrishankar S, Lyons L, Rafiq NM, Roczniak-Ferguson A, De Camilli P, Ferguson SM (2021) Overlapping roles of JIP3 and JIP4 in promoting axonal transport of lysosomes in human iPSC-derived neurons. Mol Biol Cell 32(11):1094–1103. https://doi.org/10.1091/mbc.E20-06-0382
13. Barral DC, Staiano L, Guimas Almeida C, Cutler DF, Eden ER, Futter CE, Galione A, Marques ARA, Medina DL, Napolitano G,

Settembre C, Vieira OV, Aerts J, Atakpa-Adaji-P, Bruno G, Capuozzo A, De Leonibus E, Di Malta C, Escrevente C, Esposito A, Grumati P, Hall MJ, Teodoro RO, Lopes SS, Luzio JP, Monfregola J, Montefusco S, Platt FM, Polishchuck R, De Risi M, Sambri I, Soldati C, Seabra MC (2022) Current methods to analyze lysosome morphology, positioning, motility and function. Traffic 23(5):238–269. https://doi.org/10.1111/tra.12839

14. Chazotte B (2011) Labeling lysosomes in live cells with LysoTracker. Cold Spring Harb Protoc 2011(2):pdb prot5571. https://doi.org/10.1101/pdb.prot5571
15. Chikte S, Panchal N, Warnes G (2014) Use of LysoTracker dyes: a flow cytometric study of autophagy. Cytometry A 85(2):169–178. https://doi.org/10.1002/cyto.a.22312
16. Lenk GM, Park YN, Lemons R, Flynn E, Plank M, Frei CM, Davis MJ, Gregorka B, Swanson JA, Meisler MH, Kitzman JO (2019) CRISPR knockout screen implicates three genes in lysosome function. Sci Rep 9(1):9609. https://doi.org/10.1038/s41598-019-45939-w
17. Yin Q, Jian Y, Xu M, Huang X, Wang N, Liu Z, Li Q, Li J, Zhou H, Xu L, Wang Y, Yang C (2020) CDK4/6 regulate lysosome biogenesis through TFEB/TFE3. J Cell Biol 219(8). https://doi.org/10.1083/jcb.201911036
18. Tian R, Abarientos A, Hong J, Hashemi SH, Yan R, Drager N, Leng K, Nalls MA, Singleton AB, Xu K, Faghri F, Kampmann M (2021) Genome-wide CRISPRi/a screens in human neurons link lysosomal failure to ferroptosis. Nat Neurosci 24(7):1020–1034. https://doi.org/10.1038/s41593-021-00862-0
19. Wasnick RM, Shalashova I, Wilhelm J, Khadim A, Schmidt N, Hackstein H, Hecker A, Hoetzenecker K, Seeger W, Bellusci S, El Agha E, Ruppert C, Guenther A (2022) Differential LysoTracker uptake defines two populations of distal epithelial cells in idiopathic pulmonary fibrosis. Cells 11(2). https://doi.org/10.3390/cells11020235
20. Lin HJ, Szmacinski H, Lakowicz JR (1999) Lifetime-based pH sensors: indicators for acidic environments. Anal Biochem 269(1):162–167. https://doi.org/10.1006/abio.1999.4011
21. Mustaly-Kalimi S, Gallegos W, Marr RA, Gilman-Sachs A, Peterson DA, Sekler I, Stutzmann GE (2022) Protein mishandling and impaired lysosomal proteolysis generated through calcium dysregulation in Alzheimer's disease. Proc Natl Acad Sci USA 119(49):e2211999119. https://doi.org/10.1073/pnas.2211999119
22. Sanyal A, Novis HS, Gasser E, Lin S, LaVoie MJ (2020) LRRK2 kinase inhibition rescues deficits in lysosome function due to heterozygous GBA1 expression in human iPSC-derived neurons. Front Neurosci 14:442. https://doi.org/10.3389/fnins.2020.00442
23. Somogyi A, Kirkham ED, Lloyd-Evans E, Winston J, Allen ND, Mackrill JJ, Anderson KE, Hawkins PT, Gardiner SE, Waller-Evans H, Sims R, Boland B, O'Neill C (2023) The synthetic TRPML1 agonist ML-SA1 rescues Alzheimer-related alterations of the endosomal-autophagic-lysosomal system. J Cell Sci 136(6). https://doi.org/10.1242/jcs.259875
24. Fernandopulle MS, Prestil R, Grunseich C, Wang C, Gan L, Ward ME (2018) Transcription factor-mediated differentiation of human iPSCs into neurons. Curr Protoc Cell Biol 79(1):e51. https://doi.org/10.1002/cpcb.51

Chapter 4

Label-Free Imaging of Lysosomes in Living Cells by Photothermal Microscopy

Jun Miyazaki and Ryotaro Wagatsuma

Abstract

Photothermal microscopy is an optical imaging technique used to visualize the distribution of trace amounts of endogenous dyes in living cells. This technique facilitates lysosomal observation and functional analysis by detecting and identifying accumulated substances within them, eliminating the need for molecular labeling. Here, we describe the fundamental principles and system configuration of a photothermal microscope. We also present a detailed procedure for performing photothermal measurements in living cells.

Key words Photothermal microscope, Live-cell imaging, Label-free, Optical absorption, Autophagy

1 Introduction

The photothermal (PT) effect refers to the process in which a substance absorbs light energy and converts it into heat. PT microscopy exploits this phenomenon to detect and visualize the distribution of light-absorbing molecules with high sensitivity and spatial resolution [1, 2]. Unlike fluorescent microscopy, which relies on fluorescence molecules as probes, PT microscopy enables the detection of nonfluorescent molecules that primarily absorb light without emitting fluorescence. This technique allows for the imaging of trace amounts of endogenous dyes within cells, facilitating the visualization of organelles such as lysosomes and mitochondria without the need for labeling [3–5]. By offering a distinct imaging modality from fluorescence-based techniques, PT microscopy provides valuable insights into the functions of these organelles and their roles in cellular processes.

Within lysosomes, transported substances undergo degradation, and their degradation products absorb light, generating PT effects. The intensity of the PT signal in lysosomes correlates with their degradation capacity and autophagic activity [6]. Therefore,

Stephanie M. Hughes and Indranil Basak (eds.), *Lysosomes: Methods and Protocols*, Methods in Molecular Biology, vol. 2976, https://doi.org/10.1007/978-1-0716-4844-5_4,

PT microscopy facilitates the investigation of lysosomal degradation function in living cells. It also enables the analysis of lysosomal content, as demonstrated by the detection of ferritin cluster accumulation under glucose deprivation conditions. Although PT microscopy is less established than fluorescence microscopy and its application in cellular imaging remains limited, it holds great potential for uncovering lysosomal functions. This technique could contribute to the diagnosis and understanding of diseases related to lysosomal dysfunction, including lysosomal storage disorders.

2 Principle of PT Microscopy

When light interacts with molecules, it excites them to higher electronic and vibrational energy states. As these excited molecules return to their ground state, they release the absorbed energy. Fluorescent molecules primarily emit this energy as light (fluorescence), whereas nonfluorescent molecules with low fluorescence quantum yields dissipate most of their absorbed energy as heat. This heat increases the local temperature, thereby creating a temperature gradient in the surrounding medium. As the temperature increases, the refractive index of the surrounding medium generally decreases owing to thermal expansion. For example, in water, a 1 K rise in temperature results in a refractive index decrease of 10^{-4}. Consequently, the PT effect alters the refractive index around the molecules, forming a thermal lens that diffracts the passing light in a manner similar to a conventional lens.

In a PT microscope, two laser beams, referred to as the pump and probe beams, are directed at the sample. The pump beam induces PT effects, creating a thermal lens in the sample (Fig. 1). This lens causes the probe beam to refract and changes its transmitted intensity. The magnitude of this intensity change is directly proportional to the absorbance or concentration of light-absorbing molecules in the sample. By scanning the laser across the sample and measuring the intensity changes, an image can be reconstructed that reflects the spatial distribution of molecular concentration in the sample.

3 System Configuration of PT Microscopy

Currently, PT microscopes are not commercially available from major manufacturers. Therefore, a custom measurement system must be assembled for cellular imaging and lysosomal analysis. Those with experience in optical experiments or expertise in electrical and optical measurement techniques can construct a PT microscope by assembling the lenses and mirrors on an optical table. Alternatively, a customized system can be developed in

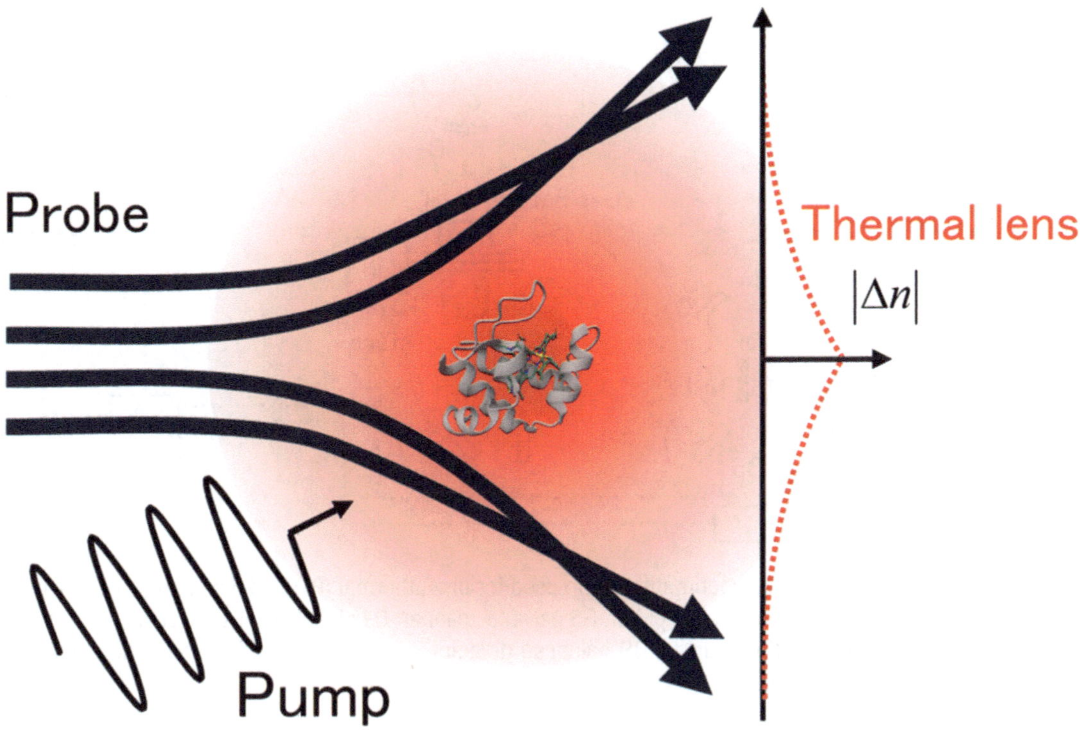

Fig. 1 Conceptual illustration explaining the working principle of photothermal microscopy

collaboration with a company or technician specializing in optical measurements. Because a PT microscope is a type of laser-scanning microscope, its basic configuration is similar to that of a confocal fluorescence microscope. Recently, optical equipment manufacturers have begun offering components, such as laser scanners for confocal microscopy, simplifying the construction of laser-scanning microscopes by integrating these components. A major advantage of building a custom microscope is the flexibility in selecting the light source and configuring the system to align it with specific research objectives.

In this section, we describe the system configuration of the dual-wavelength PT microscope developed by our group for live-cell imaging, focusing on its primary components and their operational mechanisms (Fig. 2).

3.1 Light Sources

PT microscopy typically employs lasers of different wavelengths for pump and probe beams. In our microscope, laser diodes were employed for both beams. A laser diode is a semiconductor-based light source similar to a light-emitting diode. Compared to other types of lasers, laser diodes offer several advantages: they are compact, operate at a lower voltage, are easy to handle, and are more cost-effective. Additionally, they have a long operational life of approximately 10,000 h and require minimal maintenance. The

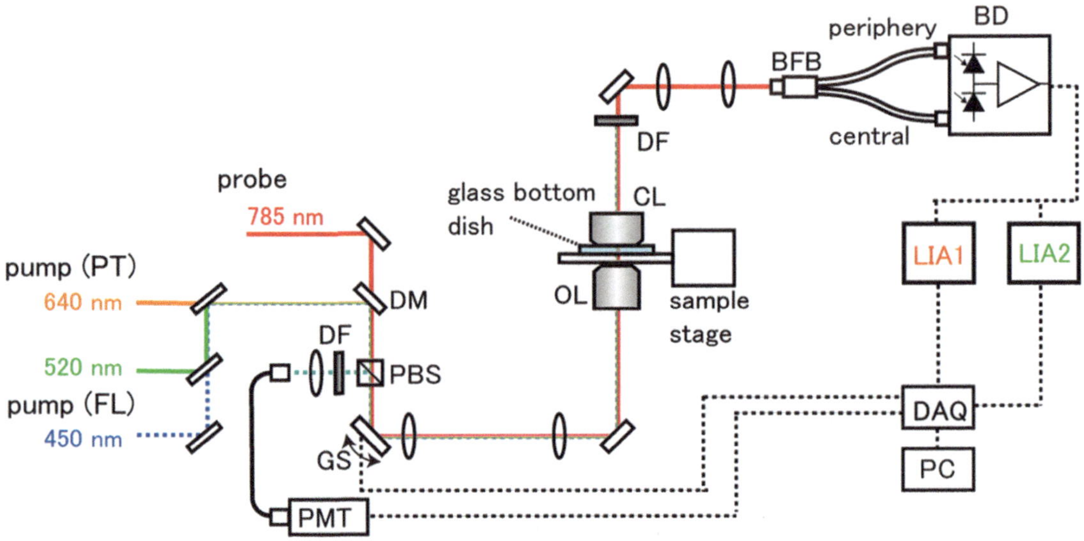

Fig. 2 Schematic diagram of the experimental setup used for photothermal (PT) and fluorescent (FL) imaging. DM dichroic mirror, PBS polarizing beam splitter, GS galvano scanner, OJ objective lens, CL collection lens, DF dichroic filter, BFB bifurcated fiber bundle, BD balanced detector, LIA lock-in amplifier, PMT photomultiplier tube, DAQ Data Acquisition device

emission wavelength is determined by the band gap of the semiconductor material, allowing laser diodes to emit across a wide spectral range, from ultraviolet to near-infrared, depending on the type of semiconductor used.

The primary organelles visualized in the cells using PT microscopy are lysosomes and mitochondria. These organelles can be distinguished by employing two pump lights of different wavelengths, as each exhibits a distinct absorption spectrum. In our setup, two laser diodes with wavelengths of 520 nm (PLP500; Osram, Germany) and 640 nm (HL6385DG; Opnext, USA) were used as the pump lights. Although the maximum output of the pump lasers ranges from 100 to 150 mW, the optical power incident on the sample was set to 4 mW at 520 nm and 15 mW at 640 nm to prevent photodamage. The temperature increase due to pump light absorption by the cells was estimated to be only a few degrees, suggesting minimal thermal effects on cellular function. For the probe light, a wavelength that minimizes the influence of light absorption and scattering within the sample is required. Therefore, a single-frequency laser diode with a wavelength of 785 nm (LP785-SAV50; Thorlabs, USA) was selected. The optical power of the probe light incident on the samples was 9 mW. The pump and probe beams were aligned along the same optical axis using dichroic mirrors and were focused onto the same region of the sample.

3.2 Objective and Condenser Lens

In a PT microscope, the pump light is directed onto the sample through an objective lens to induce the PT effect. The subsequent formation of a thermal lens is detected using a coaxially aligned probe light. The transmitted probe light is collimated using a condenser lens. Similar to a transmission microscope, the objective and condenser lenses are positioned on either side of the sample. A dichroic filter blocks the transmitted pump beam, permitting only the probe beam to reach the light detector. An objective lens with a higher numerical aperture (NA) can focus laser light more tightly on the sample, thereby increasing the light intensity at the focal point and enhancing the PT effect. However, to achieve a high signal-to-noise ratio imaging, a condenser lens is required to collect all of the probe light transmitted through the sample. Therefore, a condenser lens with a larger NA than that of the objective lens was used. In our typical experimental setup, we used an objective lens with an NA of 0.95 (UPLSAPO40X; Olympus, Japan) or 1.25 (UPLSAPO40XS; Olympus, Japan) and a condenser lens with an NA of 1.4 (U-AAC; Olympus, Japan). For live-cell imaging, the lenses are arranged in the same configuration as that of an inverted microscope, with the objective lens positioned at the bottom and the condenser lens located at the top. Both lenses are mounted on micrometer stages, allowing for precise positional adjustments.

3.3 Signal Measurement Using a Lock-in Amplifiers

When a thermal lens is formed in the sample, it refracts the probe light, thereby altering the intensity distribution of the transmitted probe beam. However, this change in intensity is minimal, typically ranging from 10^{-3} to 10^{-4} of the relative intensity. Such a small signal is difficult to detect using a conventional voltmeter as it is masked by noise.

To overcome this limitation, a lock-in amplifier is commonly used in optical measurements to detect weak signals obscured by noise. It operates on the principle of homodyne detection, in which an alternating current (AC) signal at a specific frequency is isolated using a bandpass filter. This filter effectively removes unwanted noise while extracting only the signal at the target frequency.

In the lock-in measurements, the intensity of the pump light is modulated at a specific frequency, leading to the periodic formation and dissipation of the thermal lens in the sample. Consequently, the intensity distribution of the transmitted probe beam is also modulated at the same frequency. The lock-in amplifier then measures the amplitude of the AC signal corresponding to this frequency. In the dual-wavelength PT measurement, the intensities of the two pump lights at 520 and 640 nm are modulated at different frequencies, and two lock-in amplifiers (LI5660; NF, Japan) are used to simultaneously detect each frequency component. This technique is analogous to frequency-division multiplexing in telecommunications, where signals at different frequencies are transmitted over a single channel. In our setup, the modulation frequencies of the two pump lights were set to 400 and 450 kHz.

The bandwidth of the bandpass filter in the lock-in amplifier is adjustable. Reducing the bandwidth decreases noise but increases the measurement time. To acquire an image within a practical timeframe, we perform measurements at 10 μs per pixel and set the reciprocal of the filter bandwidth (time constant) to 10 μs. Under these conditions, a 1000 × 1000 pixel image can be obtained in approximately 12 s.

3.4 Spatial-Division Balanced Detection

Since endogenous dyes in cells are present in minimal amounts, reducing noise and improving PT signal detection efficiency are essential for achieving clear cellular imaging. To enhance the signal-to-noise ratio in PT measurements, a spatial-division balanced detection method is employed in combination with the lock-in detection [7].

The formation of a thermal lens in a sample alters the propagation of the probe beam, causing it to either diverge (spread out) or converge (focus), depending on the lens's position. If the thermal lens induces divergence in a Gaussian probe beam, the intensity distribution of the transmitted light is modified, as shown in Fig. 3a. Specifically, the intensity at the center of the transmitted

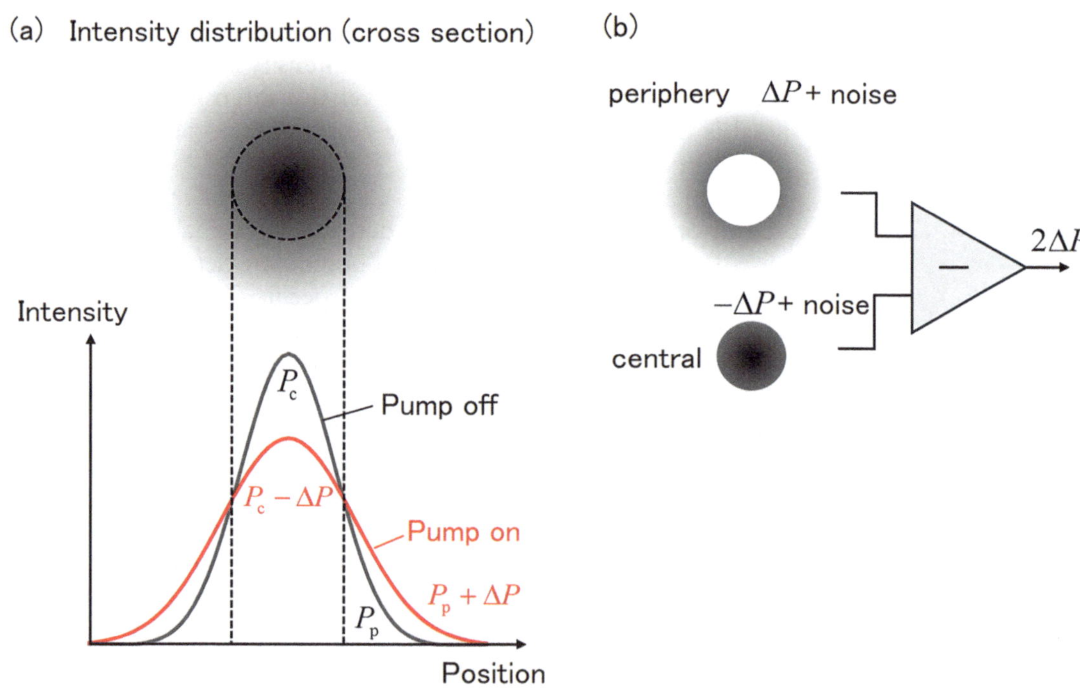

Fig. 3 Conceptual illustration of spatial-division balanced detection. (**a**) Changes in the intensity distribution of the transmitted probe beam induced by the photothermal effect. (**b**) Principle of spatial-division balanced detection

light decreases, while it increases at the periphery. However, the total power of the transmitted light remains unchanged.

The PT microscope detects changes in the intensity distribution of the probe light as signals. In spatial-division balanced detection, the transmitted probe beam is divided into central and peripheral sections, and the power difference between these sections is measured (Fig. 3b). For instance, let P_c and P_p represent the powers at the center and periphery of the probe beam, respectively, in the absence of a thermal lens. When a thermal lens forms, the power distribution shifts to $P_c - \Delta P$ and $P_p + \Delta P$, respectively. Consequently, this detection technique measures the corresponding signal difference of $2\Delta P$. This method effectively reduces the in-phase noise of the probe beam, such as noise induced by mechanical vibrations in the instruments, thereby enabling highly sensitive detection of PT-induced light intensity changes.

Our PT microscope used a bifurcated fiber bundle to separate the central and peripheral regions of the transmitted probe beam [8]. This custom-made fiber bundle was manufactured by Sumita Optical Glass (Japan). The difference in optical power at the output of the fiber bundle was detected using a balanced photodetector, which was connected to the lock-in amplifiers.

3.5 Dual Modal Imaging Combining PT and Fluorescent Measurement

Integrating a confocal fluorescence detection system into a PT microscope enables the simultaneous acquisition of both fluorescence and PT images from the same sample. The fluorescence image reveals the distribution of fluorescent molecules in the sample, whereas the PT image highlights the distribution of nonfluorescent substances. This dual imaging approach provides complementary information that facilitates a more comprehensive analysis of cellular functions.

To capture fluorescence images, a laser with an appropriate excitation wavelength for fluorescent molecules was used. Alternatively, the pump light of the PT microscope can serve as the excitation source. A polarizing beam splitter is placed in the laser path to direct the fluorescence emitted from the sample to the photomultiplier tube. After reflection from the beam splitter, the fluorescence is focused using a lens. A pinhole or optical fiber is positioned at the focal point to eliminate out-of-focus light, ensuring that only the fluorescence from the focal plane reaches the detector. Additionally, a dichroic filter is employed to block the excitation light. This confocal optical system enables the acquisition of high-contrast fluorescence images, even for thick biological tissue samples.

3.6 Laser Scanning and Data Acquisition

A two-axis Galvano scanner is used in the laser-scanning microscope to control the position of the laser beam on the sample. This device adjusts the mirror angle in response to an applied voltage, enabling precise control of the beam direction and position. To acquire the PT image, the signals from the lock-in amplifiers are recorded in synchronization with the movement of the Galvano scanner.

A simultaneous sampling board with multiple input channels (PCI-6110; NI, USA) is connected to a PC for data acquisition. The control and measurement software was developed using LabVIEW, a platform from NI.

4 Live-Cell Imaging by PT Microscopy

Assembling a PT microscope requires expertise in optoelectronics. However, once assembled, it is easy to operate and requires minimal maintenance. This simplicity allows researchers in the life sciences to easily incorporate PT microscopy into their routine research. Because PT imaging does not require molecular labeling, the sample preparation is straightforward, reducing both time and cost.

In this section, we describe a method for PT imaging of living cells. As an example, we outline a procedure for evaluating autophagic activity in cervical cancer cells (HeLa) using PT imaging. For comparison, fluorescence imaging using an autophagy detection dye was simultaneously conducted. If only PT imaging is conducted, the fluorescence labeling and observation steps can be omitted.

4.1 Sample Preparation

1. Seed the cells in 50-mm glass-bottom dishes (Matsunami, Japan) and culture them in 3 mL of Dulbecco's modified Eagle medium (DMEM) at 37 °C with 5% CO_2 for approximately 24 h.
2. Stain the cells with an autophagy detection fluorescent probe (DAPGreen; Dojindo, Japan) according to the manufacturer's instructions. DAPGreen specifically labels autophagosomes and autolysosomes. To prevent interference with PT imaging at 520 and 640 nm, a fluorescent probe with a peak excitation wavelength between 425 and 475 nm is selected.
3. Induce autophagy by replacing the medium in the glass-bottom dish with serum-free DMEM. Since phenol red, commonly included in DMEM as a pH indicator, absorbs light below 600 nm and may generate unwanted PT signals, the culture medium was replaced with phenol red-free DMEM before measurement.

4.2 Operation of the PT Microscope

1. Turn on the laser and other measuring instruments. The laser may become unstable immediately after the oscillation begins; thus, allow it to stabilize for at least 30 min to ensure the light intensity reaches a steady level.
2. Place the glass-bottom dish with cultured cells on the sample stage. The position of the condenser lens should be adjusted

such that its tip remains immersed in the culture medium. While continuously capturing low-resolution PT images (300 × 300 pixels), fine-tune the position of the glass-bottom dish to obtain an optimal image. During this adjustment, use a shutter to ensure that the laser light irradiates the sample only during image acquisition, thus minimizing photodamage.

3. Capture high-resolution PT images. Since PT measurements are unaffected by ambient light, imaging can be performed with room lights on. Adjust the pixel size according to the observation area. For example, when imaging a 100 × 100 μm area, set the resolution to 1000 × 1000 pixels. At this resolution, image acquisition takes 12 s.
4. Capture the fluorescence image by turning on the photomultiplier tube and illuminating the sample with a 450 nm laser. Use shutters or light blockers to prevent any unused laser from reaching the sample. To minimize stray light interference, ensure that the room lights are turned off.
5. Replace the samples for additional measurements. First, remove the condenser lens and then replace the glass-bottom dish. After completing observations, thoroughly wash the glass-bottom dish so that it can be reused for future measurements.

4.3 Image Analysis

Image analysis is performed using NIH ImageJ [9]. A color image is generated by assigning the images obtained at pump wavelengths of 520 and 640 nm to the green and red channels, respectively. In the resulting two-color image, lysosomes appear as brown or yellow punctate structures, while mitochondria appear as green filamentous structures (Fig. 4).

The PT signal intensity of lysosomes increases when autophagy is induced by serum starvation (Fig. 5), likely due to the accumulation of degradation residues within lysosomes. In addition, the fluorescence signal from DAPgreen also increases. The co-localization of the PT signal with the fluorescent signal confirms that the PT signal is useful for assessing autophagic activity. To quantify autophagic activity using PT images, the images are segmented to distinguish lysosomes from mitochondria based on their color and signal intensity. The PT signal intensities of the pixels corresponding to lysosomes and mitochondria are then summed for each cell. Finally, the ratio of these values (Lyso/Mito) is calculated and used as a quantitative indicator of autophagic activity.

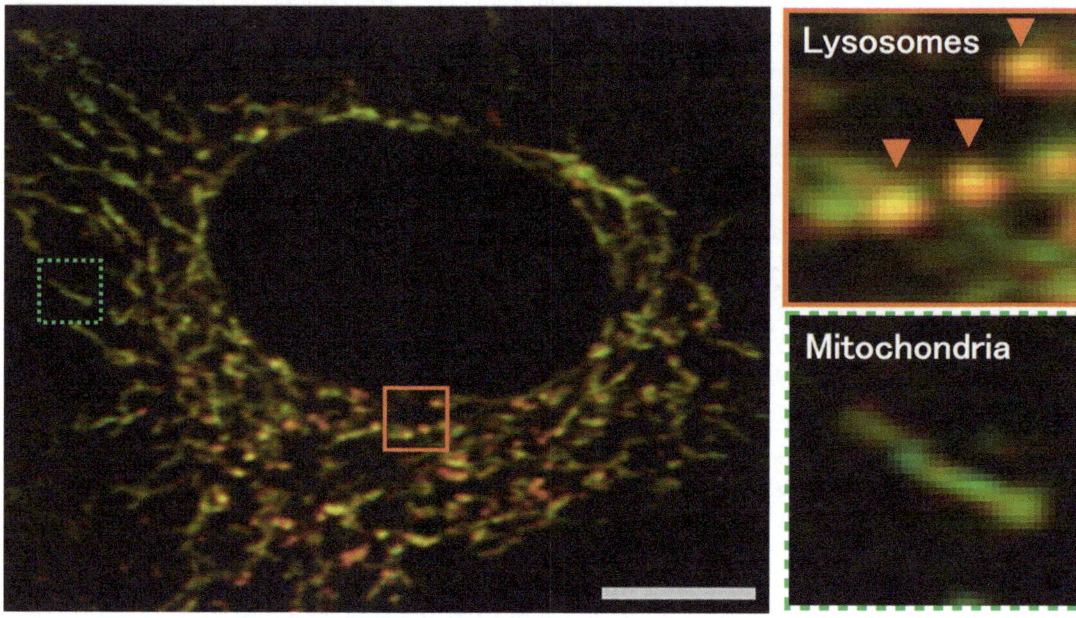

Fig. 4 Photothermal image of a live HeLa cell. Scale bar: 10 μm

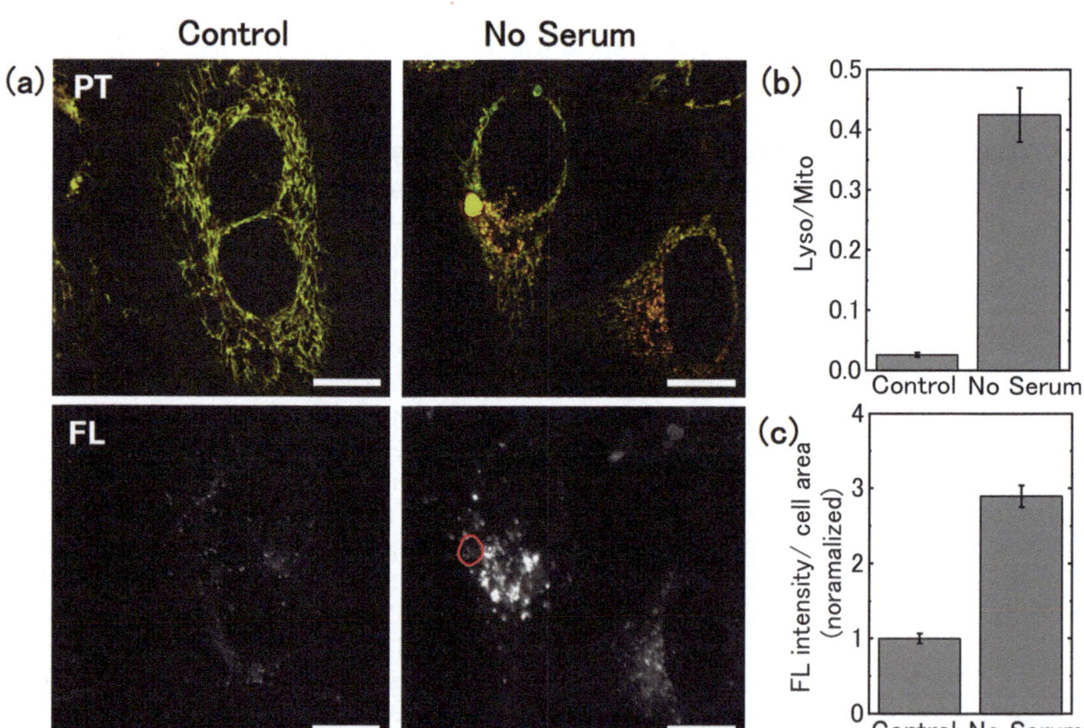

Fig. 5 Relationship between lysosomal photothermal (PT) signal and autophagic activity under serum starvation conditions. (**a**) PT and fluorescent (FL) images of cells stained with DAPGreen. Cells were cultured in serum-free medium for 1 day. Scale bar: 20 μm. (**b**) Averages of the Lyso/Mito ratio from 58 (control) and 36 (no serum) cells, with the bars representing the standard errors. (**c**) Evaluation of the fluorescence signal from the cells, with the bars representing the standard error of the mean. (Reproduced from Ref. [6] with permission from the Royal Society of Chemistry)

References

1. Adhikari S, Spaeth P, Kar A, Baaske MD, Khatua S, Orrit M (2020) Photothermal microscopy: imaging the optical absorption of single nanoparticles and single molecules. ACS Nano 14:16414–16445
2. Shimizu H, Chen CH, Tsuyama Y, Tsukahara T, Kitamori T (2022) Photothermal spectroscopy and micro/nanofluidics. J Appl Phys 132(6): 060902
3. Tamaki E, Sato K, Tokeshi M, Sato K, Aihara M, Kitamori T (2002) Single-cell analysis by a scanning thermal lens microscope with a microchip: direct monitoring of cytochrome c distribution during apoptosis process. Anal Chem 74(7): 1560–1564
4. Lasne D, Blab GA, De Giorgi F, Ichas F, Lounis B, Cognet L (2007) Label-free optical imaging of mitochondria in live cells. Opt Express 15(21):14184–14193
5. Miyazaki J, Toumon Y (2019) Label-free dynamic imaging of mitochondria and lysosomes within living cells via simultaneous dual-pump photothermal microscopy. Biomed Opt Express 10(11):5852–5861
6. Miyazaki J, Wagatsuma R, Okamoto K (2025) Photothermal imaging of cellular responses to glucose deprivation. RSC Chem Biol 6. https://doi.org/10.1039/d4cb00269e
7. Miyazaki J, Tsurui H, Kawasumi K, Kobayashi T (2015) Sensitivity enhancement of photothermal microscopy with radially segmented balanced detection. Opt Lett 40(4):479–482
8. Miyazaki J, Iida T, Tanaka S, Hayashi-Takagi A, Kasai H, Okabe S, Kobayashi T (2016) Fast 3D visualization of endogenous brain signals with high-sensitivity laser scanning photothermal microscopy. Biomed Opt Express 7(5): 1702–1710
9. Schneider CA, Rasband WS, Eliceiri KW (2012) NIH image to ImageJ: 25 years of image analysis. Nat Methods 9(7):671–675

Chapter 5

Monitoring Autophagosome Biogenesis and Degradation in Mammalian Cells Using LC3B Blotting

Jennifer E. Palmer and David C. Rubinsztein

Abstract

Autophagy is a conserved lysosomal degradation pathway that recycles protein aggregates and damaged organelles to maintain cytoplasmic quality control. Measuring the amount of the lipid-conjugated autophagic protein LC3B-II is a useful way to test whether a particular perturbation affects autophagy. However, the level of LC3B-II is affected by factors that alter either the rate of autophagosome biogenesis or degradation. Consequently, the same steady-state LC3B-II level can be reached by opposing autophagic fluxes. It is thus essential when measuring LC3B-II to perform the assay both in the absence and presence of a lysosomal inhibitor, enabling measurement of the rate of synthesis independent of its degradation. LC3B-II is also a small protein that can be challenging to detect by western blotting. In this chapter, we will provide a method for the efficient western blotting of LC3B-II and guidance as to the interpretation of the results.

Key words Autophagy, LC3B, Western blotting, SDS-PAGE, ATG8, Bafilomycin A_1

1 Introduction

Autophagy is an essential, conserved cellular recycling pathway that is critically dependent on efficient lysosomal function. Autophagic cargos, such as aggregated proteins and damaged organelles, are packaged into double-membraned autophagosomes. Autophagosomes then fuse with lysosomes to form autolysosomes, where the autophagic cargo is degraded by lysosomal hydrolases. Autophagy is an important cytoplasmic quality control route, and impairments in its function have been implicated in a wide variety of diseases [1–5].

Mammalian autophagy can be monitored using the critical autophagic marker protein LC3B, which is ubiquitin-like. Nascently translated LC3B (pro-LC3B) is cleaved at its C-terminus by the ATG4 protease to form LC3B-I, which has an exposed C-terminal glycine, like ubiquitin. This glycine in LC3B-I is then

Stephanie M. Hughes and Indranil Basak (eds.), *Lysosomes: Methods and Protocols*, Methods in Molecular Biology, vol. 2976, https://doi.org/10.1007/978-1-0716-4844-5_5,

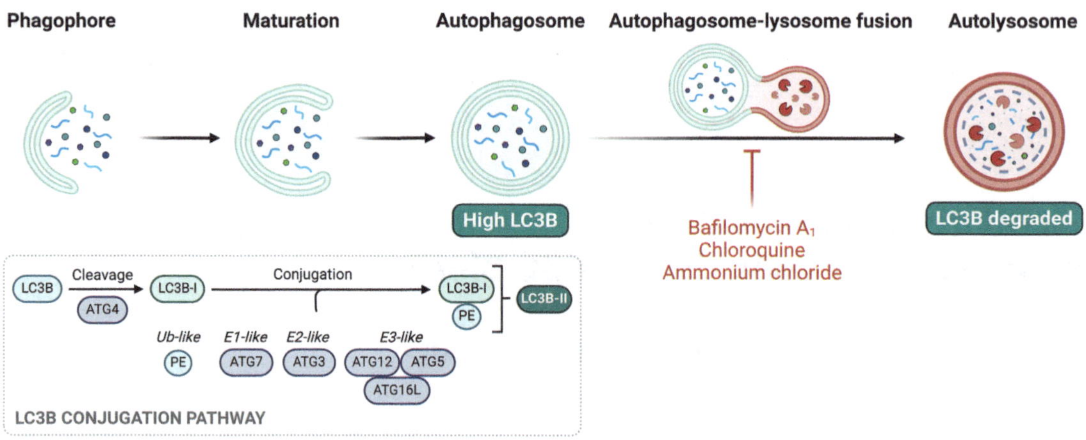

Fig. 1 LC3B-II is a marker of the number or volume of autophagosomes per cell. LC3B is cleaved into LC3B-I by the ATG4 protease. ATG12 (as another ubiquitin-like molecule in this pathway) is conjugated to ATG5, and then these form a complex with ATG16L1. Then the ATG12–5/16L1 complex serves as an E3-like ligase to enable LC3B conjugation to phosphatidylethanolamine (PE) in nascent autophagosome membranes to form LC3B-II. LC3B-II is incorporated into the inner and outer autophagosomal membranes and serves as a marker of the number of autophagosomes. LC3B-II within the autophagosome is degraded in the autolysosome, while LC3B-II is deconjugated from the cytoplasmic face of autophagosome-derived membranes after autolysosome formation. Lysosomal inhibitors such as bafilomycin A_1 block the lysosomal degradation of LC3B-II

conjugated to phosphatidylethanolamine (PE) in nascent autophagosome membranes by the E1-like enzyme ATG7, the E2-like ATG3, and the E3-like ATG12-5-16L1 complex, producing LC3B-II. Both LC3B-I and LC3B-II are detected with western blotting for LC3B. The pre-cleavage LC3B is not detected by western blotting under normal conditions because ATG4 cleaves LC3B immediately after translation. Despite the increased mass due to the addition of PE, LC3B-II migrates at a smaller apparent molecular weight (~14 kDa) than LC3B-I (~16 kDa), due to the hydrophobicity of the PE moiety. LC3B-II is associated with the inner and outer autophagosomal membranes and, as discovered by Yoshimori and colleagues, serves as a marker of autophagosomes [6]. Once the outer membrane of the autophagosome fuses with a lysosome, LC3B-II on the inner autophagic membrane is degraded by lysosomal hydrolases, and LC3B-II on the outer membrane is deconjugated by ATG4 and thereby released from the autolysosomes. Thus, the amount of LC3B-II is a proxy for the volume or number of autophagosomes per cell and can be used to monitor autophagy [6–8]. The process of LC3B conjugation is shown in Fig. 1.

The LC3B-II level at any given point in time only provides a snapshot of the current volume or number of autophagosomes and not the flux through the pathway. Both changes in the rate of synthesis and in the rate of degradation of LC3B-II affect the amount of LC3B-II at any given moment, a critical observation that was made by Kominami and colleagues [7]. For example, a high level of LC3B-II results if the rate of LC3B-II formation exceeds the rate of degradation. This can occur when there is increased autophagosome biogenesis (which enhances autophagic substrate clearance) or when LC3B-II degradation is compromised (which impairs autophagic substrate clearance). This is particularly relevant when considering interventions that are likely to alter lysosomal function, as researchers who are not careful when interpreting their results may erroneously conclude that their intervention has increased autophagic flux due to an increased amount of LC3B-II, when their intervention has reduced the lysosomal degradation of LC3B-II and, as a consequence, impaired autophagic flux. Note that decreased LC3B-II turnover resulting in increased steady-state levels can occur if there are blocks at any stage after autophagosome formation—including impaired autophagosome closure [9], impaired scission of nascent autophagosomes from the recycling endosome [10], and compromised transport of autophagosomes to the microtubule organizing center where lysosomes are concentrated [11], in addition to the more commonly considered possibilities of decreased autophagosome–lysosome fusion [12] or compromised lysosome function [13].

To differentiate effects on autophagosome biogenesis and degradation, western blotting for LC3B-II is performed under both basal conditions and in the presence of a lysosomal inhibitor such as bafilomycin A_1. In the presence of a lysosomal inhibitor, autophagosomes cannot be degraded, resulting in an accumulation of LC3B-II. Thus, any difference in the amount of LC3B-II in the presence of bafilomycin A_1 indicates changes in the rate of autophagosome biogenesis independent of the rate of degradation. In contrast, in the absence of a lysosomal inhibitor, increased LC3B-II could indicate either enhanced synthesis or impaired degradation, whereas decreased LC3B-II could indicate either impaired synthesis or enhanced degradation. Examples of possible scenarios and how they appear on an LC3B blot are shown in Fig. 2.

It is worth noting that an assumption of LC3B blotting is that the amount of LC3B-II per autophagosome is constant, such that a change in the amount of LC3B-II on a western blot reflects a change of the same magnitude in the number of autophagosomes. This assumption has, to the best of our knowledge, never been tested. Thus, we prefer to consider that LC3B-II levels relate to the volume, rather than the number of autophagosomes, per cell. It is plausible to hypothesize that some interventions might cause a greater increase in the LC3B-II level than in the number of autophagosomes, such as an intervention that results in larger autophagosomes. Thus, this assumption must be kept in mind when interpreting LC3B blots, and microscopy experiments should be

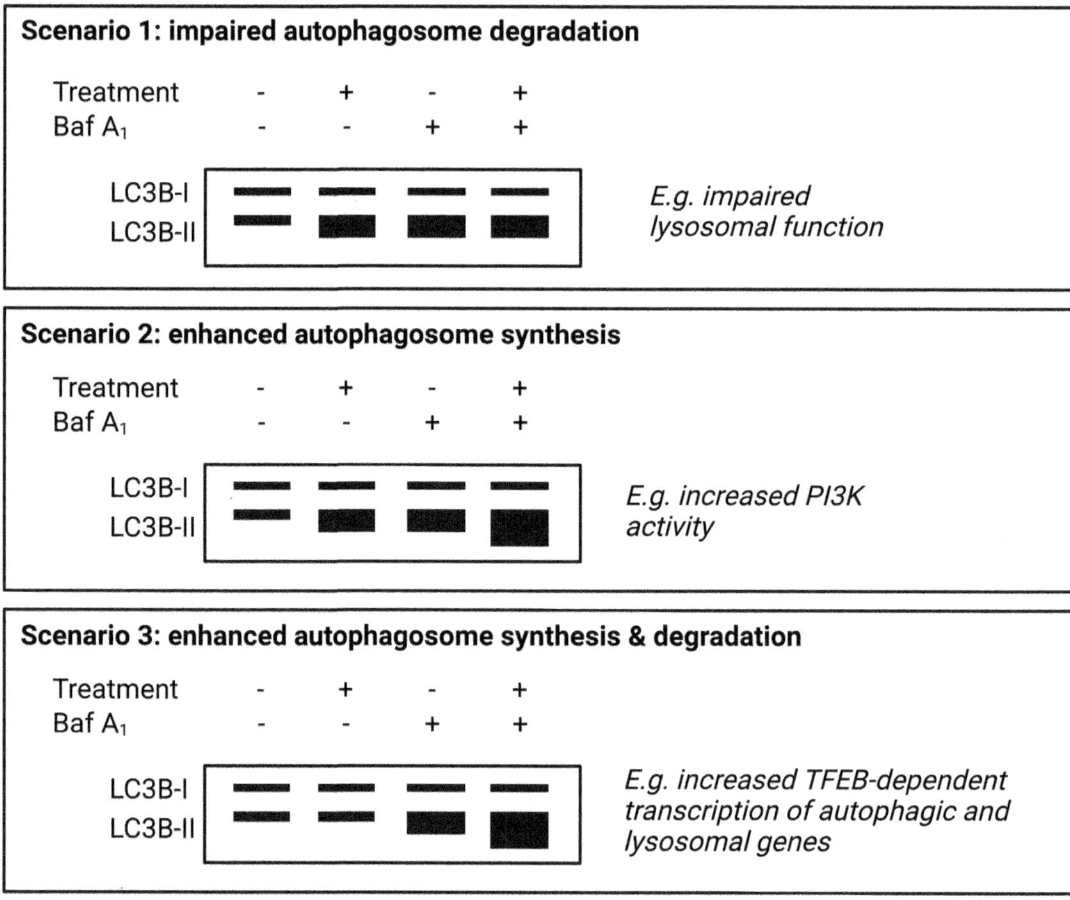

Fig. 2 Potential scenarios arising from LC3B blotting and what they indicate about the rate of autophagosome synthesis and degradation. In scenario 1, basal LC3B-II is increased, but there is no difference between the treatments in the presence of the lysosomal inhibitor bafilomycin A_1. In this scenario, there is no change to the level of autophagosome synthesis; however, the rate of autophagic degradation is impaired. This represents a decrease in autophagic flux, such as that arising from impaired lysosomal function. In scenario 2, LC3B-II is increased both in the presence and absence of bafilomycin A_1. This indicates that the rate of autophagosome biogenesis is increased, enhancing autophagic flux. Thus, the use of the lysosomal inhibitor bafilomycin A_1 enables the differentiation of a situation where autophagic degradation is impaired (scenario 1) and where autophagosome production is increased (scenario 2), despite their steady state amount of LC3B-II being the same. In scenario 3, the rate of autophagosome biogenesis is enhanced (shown by the increase in LC3B-II in the presence of bafilomycin A_1), but the steady-state rate is unaltered. This indicates that both the production of autophagosomes and the rate of their clearance have been increased. This effect could hypothetically be observed when TFEB activity is increased, because there will be increased transcription of both autophagic genes and lysosomal proteases, resulting in a large increase in autophagic flux. Indeed, it is sometimes observed in such scenarios that the basal level of LC3B-II might even be decreased due to more efficient lysosomal degradation of LC3B-II

undertaken to confirm these changes, especially in cases where the size of autophagosomes may be altered.

Another important caveat with LC3B-II measurements is that more recent studies have shown that LC3B-II can also be conjugated to single membranes under certain conditions, termed Conjugation of ATG8 to Single Membranes (CASM) [14]. For example, LC3 family members (including LC3B) can be conjugated to phagophores (termed LC3-associated phagocytosis, LAP), endosomes (LC3-associated endocytosis, LANDO), and secretory vesicles (LC3-dependent extracellular vesicle loading and secretion, LDELS) [15–20]. Thus, when making inferences about certain perturbations that may change autophagy, LC3B-II blotting should be used in conjunction with other methods that allow one to either show that the structures are double-membraned (e.g. by electron microscopy), or that they clear conventional autophagy substrates (in wild-type but not autophagy-null cells), or that the effects are blocked in the presence of interventions that impair autophagy but not CASM (e.g., ULK1/2 inhibition or knockdown) [17–19]. While LC3B is one of six human ATG8 members and the other proteins in this family are also conjugated to PE, they may not all act at the same stages of the autophagy pathway as LC3B. Thus, it is safer in the first instance to study LC3B with these assays.

In this chapter, we will provide a protocol for the effective measurement of LC3B-II levels by western blotting and guidance on the interpretation of these results. We will provide recommendations as to the use of the lysosomal inhibitor bafilomycin A_1 to differentiate between effects on autophagosome biogenesis and degradation, a protocol for the successful separation of LC3B-I and LC3B-II by SDS-PAGE, western blotting to quantify the amount of LC3B-II, and a detailed description of how to interpret such results.

2 Materials

2.1 Samples, Treatments, and Lysis

1. Cells in culture. We routinely use a variety of immortalized or primary cell lines (*see* **Note 1**). Include two wells in each condition being investigated, because one well will be treated with bafilomycin A_1 and the other with a DMSO negative control.
2. Bafilomycin A_1 stock solution (100 μM) in DMSO (*see* **Note 2**).
3. Phosphate-buffered saline (PBS) buffer at pH 7.4: 138 mM NaCl, 2.7 mM KCl, 10 mM Na_2HPO_4, 1.76 mM KH_2PO_4.
4. Radioimmunoprecipitation assay (RIPA) buffer: 150 nM NaCl, 1% NP-40, 0.5% sodium deoxycholate (NaDoC), 0.1%

sodium dodecyl sulfate (SDS), 50 mM Tris pH 7.4, 1× protease inhibitor cocktail.

5. A standard protein assay kit to quantify the total protein concentration in each sample, such as a bicinchoninic acid (BCA) assay kit or a Bradford assay kit.
6. 4× Laemmli buffer: 130 mM Tris–HCl pH 6.8, 50% (w/v) glycerol, 4% SDS, 0.02% (w/v) bromophenol blue, 10% (v/v) β-mercaptoethanol.
7. 1× Laemmli buffer: dilute the 4× buffer above 1:3 in dH_2O.

2.2 SDS-PAGE and Western Transfer

1. 30% acrylamide/bis-acrylamide solution.
2. 1.5 M Tris-HCl at pH 8.8 (for the resolving gel buffer).
3. 1.0 M Tris-HCl pH 6.8 (for the stacking gel buffer).
4. 10% SDS in dH_2O.
5. 10% ammonium persulfate in dH_2O.
6. N,N.N′,N′-tetramethylethylenediamine (TEMED).
7. dH_2O.
8. Pre-stained molecular weight markers.
9. 10× Tris-glycine buffer: 250 mM Tris pH 8.3, 1.92 M glycine.
10. Running buffer: 100 mL of 10× Tris–glycine buffer, 10 mL of 10% SDS, dH_2O to 1 L.
11. Transfer buffer: 100 mL of 10× Tris–glycine buffer, 200 mL of methanol, dH_2O to 1 L.
12. SDS-PAGE apparatus including a tank, glass plates, casting cassettes, gel combs, and a voltage pack.
13. Wet transfer western blotting apparatus, including clean sponges.
14. 0.45-μm PVDF western blotting membrane.
15. Methanol (HPLC grade).

2.3 Antibody Incubations and Detection

1. PBS–Tween: 0.1% Tween-20 in PBS.
2. Blocking buffer: lean (skim) milk powder 5% (w/v) in PBS–Tween.
3. LC3B antibody. We use a ProteinTech polyclonal rabbit LC3B antibody (14600-1-AP) or an Abcam rabbit monoclonal LC3B antibody (EPR18709, ab192890) (*see* **Note 3**).
4. An appropriate loading control antibody, e.g., GAPDH, actin, or tubulin (*see* **Note 4**).
5. Secondary antibodies appropriate for your detection system. We use a fluorescently conjugated IR-dye system (*see* **Note 5**).

6. An imaging system appropriate for your secondary antibody. We use a LICOR Odyssey imager.

3 Methods

Please note that the method below is for cells grown in culture conditions. If one wishes to perform this assay on tissue samples, please make appropriate changes to the sample treatments, processing, and collection.

3.1 Cell Culture and Treatments

1. Grow cells to 60–80% confluency. It is essential that cells are not overconfluent and that the confluency is comparable between the experimental conditions (*see* **Note 6**). Two wells are required for each manipulation: one under basal conditions and one with bafilomycin A_1.
2. Treat with bafilomycin A_1 (400 nM) or an equivalent volume of DMSO for 4 h (*see* **Notes** 7 and **8**).
3. Collect and lyse cells in an appropriate volume of RIPA buffer (*see* **Note 9**).
4. Perform a BCA assay to determine the protein concentration in each sample (*see* **Note 10**).
5. Dilute samples in dH_2O and add 4× Laemmli buffer at a 1:3 ratio to the total volume of each sample.
6. Boil the samples at 100 °C for 7–10 min.

3.2 SDS-PAGE and Western Transfer

1. Prepare 15% resolving gels: for six gels, combine 12.5 mL of 1.5 M Tris-HCl pH 8.8, 12 mL of dH_2O, 24 mL of 30% acrylamide, and 500 μL of 10% SDS. Mix well, then add 400 μL of 10% ammonium persulfate and 40 μL of TEMED. Pour into a gel casting apparatus, leaving sufficient space for the stacking gel. Cover the surface with dH_2O. Leave any excess solution in a Falcon tube to show when polymerization has occurred (*see* **Note 11**).
2. Prepare 4% stacking gels: pour off the excess water from the resolving gel. Then, for 6 gels, combine 2.5 mL of 1 M Tris-HCl pH 6.8, 13.6 mL of dH_2O, 3.4 mL of 30% acrylamide, and 200 μL of 10% SDS. Mix well, then add 200 μL of 10% ammonium persulfate and 20 μL of TEMED. Pipette carefully, ensuring not to create any bubbles. Insert a gel comb with the desired number of wells. Leave any excess solution in a Falcon tube to show when polymerization has occurred.
3. Move the polymerized gels into an SDS-PAGE apparatus and remove the combs. Fill with running buffer.

4. Load up to 20 μg of total protein per lane (*see* **Notes 12** and **13**).
5. Load a molecular weight marker in one of the empty lanes to enable confirmation of protein sizes.
6. Run at 80 V through the stacking gel, then 100 V until the dye front reaches the bottom of the gel (*see* **Note 14**).
7. Soak PVDF membranes in methanol for 10–30 s, then equilibrate in transfer buffer for ~5 min.
8. Soak filter paper in transfer buffer for >5 min. Simultaneously, soak gels in transfer buffer for >5 min.
9. Set up a wet transfer western apparatus using a sponge, three pieces of filter paper, the PVDF membrane, the SDS-PAGE gel, a three further pieces of filter paper, then another sponge.
10. Transfer at 70 V for 1 h. Ensure that the transfer does not overheat by placing the apparatus in ice (*see* **Note 15**).

3.3 Antibody Blotting and Detection

1. Block in blocking buffer for 30 min at room temperature on a shaker.
2. Incubate with the LC3B primary antibody overnight at 4 °C on a shaker. For the antibodies that we use, a 1:10000 dilution in blocking buffer gives optimal staining.
3. Wash three times for 5 min each in PBS–Tween.
4. Incubate with secondary antibody for 1 h at room temperature on a shaker. Protect fluorescent antibodies from light. For the antibodies that we use, a 1:5000 dilution in blocking buffer gives optimal staining.
5. Wash three times for five minutes each in PBS–Tween.
6. Image, taking care not to overexpose the membrane and to image within the linear range of detection (*see* **Note 16**).

3.4 Quantification and Interpretation of the Results

1. Quantify the intensity of the LC3B-II band in each lane. LC3B-II is the smaller migrating LC3B band (*see* Fig. 2). We use the Image Studio software. Use the same-sized rectangular box for each band to ensure that any differences in the background intensity do not affect the quantification (*see* **Notes 16** and **17**).
2. Quantify the intensity of the loading control band in each lane.
3. Divide the intensity of the LC3B-II band by that of the loading control band for each lane (*see* **Note 18**).
4. Normalize the LC3B-II/loading control ratio by dividing it by the ratio in the control sample to obtain a fold change in the normalized amount of LC3B-II for each sample. It is common to perform a separate normalization for samples in the presence and absence of lysosomal inhibition; this is to account for

differences in the fold change induced by lysosomal inhibition in different experiments (*see* **Note 19**).

5. To interpret what the LC3B-II levels indicate in terms of autophagic flux, please refer to Fig. 2. Autophagic flux is likely enhanced in samples that have: (a) increased LC3B-II in both basal and lysosomal inhibition conditions, indicating enhanced autophagosome biogenesis; (b) increased LC3B-II in lysosomal inhibited but not basal conditions, indicating both enhanced autophagosome biogenesis and enhanced autophagosome clearance; (c) decreased LC3B-II in basal conditions but unaltered levels in the presence of lysosomal inhibitors, indicating an enhanced rate of autophagosome degradation. Autophagic flux is likely impaired in samples that have: (a) decreased LC3B-II in both basal and lysosomal inhibition conditions, indicating impaired autophagosome biogenesis; (b) increased LC3B-II in basal conditions but not in the presence of lysosomal inhibitors, indicating a decreased rate of autophagosome degradation (*see* **Note 20**).

3.5 Follow-Up Experiments

LC3B blotting is a useful starting point for determining how a given intervention affects autophagy because, when combined with the use of lysosomal inhibitors, it enables the differentiation of effects on autophagosome biogenesis and autophagic degradation. Where appropriate, it may be necessary to discriminate effects of autophagy from those on CASM by testing if the structures are double-membraned (e.g., by electron microscopy), or the perturbations alter clearance of conventional autophagy substrates (in wild-type but not autophagy-null cells), or that the effects are blocked in the presence of interventions that impair autophagy but not CASM (e.g., ULK1/2 inhibition/knockdown).

If autophagosome biogenesis is altered, one may wish to follow up with further blots to determine whether the activity of upstream signaling complexes such as the mTORC1, ULK1, AMPK, or PI3K class III complexes is altered [21–23]. If autophagic degradation is altered, one may wish to determine whether lysosomal function is altered or the fusion of autophagosomes and lysosomes is affected [24–26]. Further assays that measure autophagic substrate clearance can be used to confirm the conclusions of the LC3B-II western blotting.

4 Notes

1. We routinely use HeLa, SH-SY5Y, HEK293T, H4, and BV2 immortalized cell lines; patient fibroblasts; primary murine neurons and glia; induced pluripotent stem cells(iPSC); and iPSC-derived neuronal cells.

2. Inhibition of the lysosomal degradation of LC3B-II is required in order to differentiate effects on LC3B-II production and degradation. Bafilomycin A_1 inhibits the lysosomal vATPase, reducing lysosomal acidity and inhibiting autophagosome-lysosomal fusion [27]. Alternate methods of inhibiting the lysosomal degradation of LC3B-II can be used. For example, ammonium chloride or chloroquine also impair lysosomal acidity, whereas pepstatin A and leupeptin inhibit lysosomal proteases. The use of chloroquine is not recommended because chloroquine can stimulate CASM, potentially confounding the results [28].
3. Under our conditions, both of the antibodies that we recommend detect LC3B by western blotting. 14600-1-AP detects mostly LC3B-II, whereas ab192890 is sensitive to both LC3B-I and LC3B-II.
4. A suitable loading control antibody is one that does not change in response to the conditions or treatment that are being used in this experiment. In our experience, the total amount of GAPDH, actin, and tubulin is unaltered by starvation or other common manipulations. However, some drugs or genetic changes might alter one or more of these proteins, so care must be taken to choose a suitable antibody for the experimental conditions being examined.
5. Horseradish peroxidase (HRP)-conjugated antibodies may also be used. However, the fluorescently labeled secondary antibodies have significant advantages, such as a larger linear range for quantification and the ability to multiplex antibodies of different species origin with different fluorescent colors.
6. Cell confluency affects autophagy and is a potential confounder of any LC3B blotting results. If the cell density is too high, contact inhibition results in impaired autophagosome biogenesis [29]. Thus, if a condition is overconfluent, a researcher may incorrectly assume that this condition impairs autophagy.
7. We recommend diluting the bafilomycin A_1 to 400 nM in sufficient volume for all of the bafilomycin A_1-treated samples, and doing the same for the DMSO-treated samples. These mixtures should be vortexed and then applied to all conditions. This ensures that all samples receive the same bafilomycin A_1 or DMSO exposure, reducing the technical variance.
8. In our experience, 400 nM bafilomycin A_1 for 4 h is sufficient to inhibit the lysosomal degradation of LC3B-II in most cells. However, we and others have observed that such a bafilomycin A_1 treatment often has little effect in iPSC-derived neurons. In these cells, it may be necessary to treat with bafilomycin A_1 for 24 h, or to use ammonium chloride instead. However, this appears to be unique to these cells, because primary neurons

show a strong response with 400 nM bafilomycin A_1. The purpose of the lysosomal inhibitor is to halt LC3B-II degradation in order to make inferences about LC3-II formation rates. Thus, it is critical that the concentration used is saturating; otherwise, an inhibitor of autophagic flux that acts downstream of LC3B-II formation may also increase LC3B-II levels in the presence of too little inhibitor (*see* **Note 19**). One can determine suitable concentrations of this inhibitor by doing a preliminary dose–response experiment in wild-type cells.

9. In our experience, 200 μL of lysis buffer per well of a 6-well plate is suitable. Other lysis buffers may be used. If the cell confluency is identical between the conditions, it is also possible to lyse the cells directly in 1× Laemmli buffer. However, this prevents the use of assays to quantify the protein concentration.
10. Alternate protein concentration assays, such as the Bradford assay, may also be performed.
11. If a large protein must also be stained on these gels, a 13% resolving gel also provides adequate separation of LC3B-I and LC3B-II. However, 12% gels and below often fail to adequately separate LC3B-I and LC3B-II. If a precast gradient gel is preferred, make sure to choose an 8–16% or greater percentage gel, because 4–12% gels often fail to provide sufficient separation between LC3B-I and LC3B-II.
12. One must optimize the amount of protein loaded to ensure that the amount of both LC3B-II and the loading control is within the linear range of detection. Loading too much protein can oversaturate the loading control (and occasionally LC3B-II, such as in the bafilomycin A_1-treated samples), whereas loading too little protein can lead to unreliable detection of low-abundance proteins, such as basal LC3B-II in some cells. To determine an appropriate protein amount in each experimental system, create a standard curve by loading varying total protein amounts and graphing the intensity of both LC3B-II and the loading control against the total amount of protein. Select a total protein amount where both LC3B-II and the loading control are within the linear range.
13. Try not to load any samples in the outermost lane of any gel. Samples in these outer lanes are the most prone to run at an unusual angle or transfer poorly, both of which impair accurate quantification. Instead, load 1× Laemmli buffer into the outermost lanes. This normalizes the SDS concentration and helps the other lanes to run straighter.
14. Take care not to run the gels too long. LC3B-II is a small protein and thus will be one of the first proteins to be lost from the bottom of the gel. Stop the gel from running before the dye front reaches the bottom of the gel.

15. In our experience, wet transfer as described here results in slightly better detection of LC3B-II compared to semi-dry transfer. However, if semi-dry transfer is preferred, transfer can be achieved using two thick (e.g., 0.83 mm) pieces of filter paper below the gel and one thick piece of filter paper above the gel. Semi-dry transfer should be performed at 18 V for 1 h.
16. It may be necessary to use different exposures for LC3B and the loading control antibody to ensure that both proteins are within the linear range of detection.
17. Take care that the quantification box does not overlap with the signal from any adjacent lanes, including the molecular weight marker. If this is too difficult, consider loading less protein or less molecular weight marker, or use larger well sizes (e.g., 10-well combs instead of 15-well).
18. The amount of LC3B-II should be normalized to the amount of the loading control in that sample, not to the amount of LC3B-I. This is because transcriptional upregulation of autophagy will increase the total LC3B levels, whereas induction of autophagy will deplete LC3-I as it is converted to LC3-II. Thus, the ratio of LC3B-II to LC3B-I is variable and less informative than a direct comparison of the total amount of LC3B-II normalized to the total protein amount. Furthermore, LC3B-II tends to be more sensitively detected by western blotting than LC3B-I [30].
19. Exclude the results from any experiments where the control sample treated with bafilomycin A_1 does not show at least a twofold increase in LC3B-II compared to the control sample treated with DMSO. This is because the interpretation of the LC3B blotting results assumes that the bafilomycin A_1 treatment was saturating and completely blocked the degradation of LC3B-II. If the bafilomycin A_1-treatment only caused a small increase in the total amount of LC3B-II, this could indicate that the treatment was insufficient to fully block the lysosomal degradation of LC3B-II, which prevents the researcher from being able to ascertain whether a change in the amount of LC3B-II was due to an altered rate of production or an altered rate of degradation.
20. This last example is the most likely to occur in scenarios where lysosomal function is impaired, and care must be taken when interpreting these results to conclude that the increased basal LC3B-II is due to a block in degradation, rather than an enhanced rate of synthesis. This means that ensuring that the lysosomal inhibition is saturating is of critical importance, because the difference between enhanced biogenesis and impaired degradation can only be differentiated when the rate

of autophagosome production can be assessed independently of the rate of degradation.

Acknowledgments

We are grateful for the support of a Cambridge Rutherford-Memorial PhD Scholarship from the Rutherford Foundation, Royal Society Te Apārangi, and the Cambridge Commonwealth European and International Trust. We are also grateful for funding from the UK Dementia Research Institute (funded by the Medical Research Council, Alzheimer's Research UK, and the Alzheimer's Society) and the NIHR Cambridge Biomedical Research Centre (grant no. NIHR203312). The views expressed are those of the authors and not necessarily those of the NIHR or the Department of Health and Social Care.

References

1. Aman Y, Schmauck-Medina T, Hansen M et al (2021) Autophagy in healthy aging and disease. Nat Aging 1:634–650
2. Deretic V (2021) Autophagy in inflammation, infection and immunometabolism. Immunity 54:437–453
3. Fleming A, Bourdenx M, Fujimaki M et al (2022) The different autophagy degradation pathways and neurodegeneration. Neuron 110:935–966
4. Kitada M, Koya D (2021) Autophagy in metabolic disease and ageing. Nat Rev Endocrinol 17:647–661
5. Klionsky DJ, Petroni G, Amaravadi RK et al (2021) Autophagy in major human diseases. EMBO J 40:e108863
6. Kabeya Y, Mizushima N, Ueno T et al (2000) LC3, a mammalian homologue of yeast Apg8p, is localized in autophagosome membranes after processing. EMBO J 19:5720–5728
7. Tanida I, Minematsu-Ikeguchi N, Ueno T, Kominami E (2005) Lysosomal turnover, but not a cellular level, of endogenous LC3 is a marker for autophagy. Autophagy 1:84–91
8. Tanida I, Ueno T, Kominami E (2004) LC3 conjugation system in mammalian autophagy. Int J Biochem Cell Biol 36:2503–2518
9. Puri C, Gratian MJ, Rubinsztein DC (2023) Mammalian autophagosomes form from finger-like phagophores. Dev Cell 58:2746–2760
10. Puri C, Manni MM, Vicinanza M et al (2020) A DNM2 centronuclear myopathy mutation reveals a link between recycling endosome scission and autophagy. Dev Cell 53:154–168
11. Ravikumar B, Acevedo-Arozena A, Imarisio A et al (2005) Dynein mutations impair autophagic clearance of aggregate-prone proteins. Nat Genet 37:771–776
12. Jiang P, Nishimura T, Sakamaki Y et al (2014) The HOPS complex mediates autophagosome-lysosome fusion through interaction with syntaxin 17. Mol Biol Cell 25:1327–1337
13. Yim WWY, Mizushima N (2020) Lysosome biology in autophagy. Cell Disc 6:6
14. Durgan J, Florey O (2022) Many roads lead to CASM: diverse stimuli of noncanonical autophagy share a unifying molecular mechanism. Sci Adv 8:eabo1274
15. Romao S, Munz C (2013) LC3-associated phagocytosis. Autophagy 10:526–528
16. Martinez J, Almendinger J, Oberst A et al (2011) Microtubule-associated protein 1 light chain 3 alpha (LC3)-associated phagocytosis is required for the efficient clearance of dead cells. PNAS 108:17396–17401
17. Martinez J, Malireddi RKS, Lu Q et al (2015) Molecular characterization of LC3-associated phagocytosis reveals distinct roles for Rubicon, NOX2 and autophagy proteins. Nat Cell Biol 17:893–906
18. Pena-Martinez C, Rickman AD, Heckmann BL (2022) Beyond autophagy: LC3-associated phagocytosis and endocytosis. Sci Adv 8:eabn1702

19. Magne J, Green DR (2022) LC3-associated endocytosis and the functions of Rubicon and ATG16L1. Sci Adv 8:eabo5600
20. Leidal AM, Debnath J (2020) LC3-dependent extracellular vesicle loading and secretion (LDELS). Autophagy 16:1162–1163
21. Kim K, Kundu M, Viollet B, Guan KL (2011) AMPK and mTOR regulate autophagy though direct phosphorylation of Ulk1. Nat Cel Biol 13(2):132–141
22. Russell RC, Tian Y, Yuan H, Park HW, Chang YY, Kim J, Kim H, Neufeld TP, Dillin A, Guan KL (2013) ULK1 induces autophagy by phosphorylating Beclin-1 and activating VPS34 lipid kinase. Nat Cell Biol 15(7):741–750
23. Russell RC, Yuan HX, Guan KL (2014) Autophagy regulation by nutrient signalling. Cell Res 24(1):42–57
24. Lieberman AP, Puertollano R, Raben N, Slaugenhaupt S, Walkley SU, Ballabio A (2012) Autophagy in lysosomal storage disorders. Autophagy 1;8(5):719–730
25. Nakamura S, Yoshimori T (2017) New insights into autophagosome-lysosome fusion. J Cell Sci 1;130(7):1209–1216
26. Yim WWY, Mizushima N (2020) Lysosome biology in autophagy. Cell Discov 11(6):6
27. Mauvezin C, Nagy P, Juhász G et al (2015) Autophagosome-lysosome fusion is independent of V-ATPase-mediated acidification. Nat Commun 6:7007
28. Liebl MP, Meister SC, Frey L et al (2022) Robust LC3B lipidation analysis by precisely adjusting autophagic flux. Sci Rep 12:79
29. Pavel M, Renna M, Park SJ et al (2018) Contact inhibition controls cell survival and proliferation via YAP/TAZ-autophagy axis. Nat Commun 9:2961
30. Mizushima N, Yoshimori T (2007) How to interpret LC3 immunoblotting. Autophagy 6: 542–545

Chapter 6

Measurement of Physiological Autophagic Flux in the Human Peripheral Blood Mononuclear Cell Pool

Leanne K. Hein, Kathryn J. Hattersley, Julien Bensalem, and Timothy J. Sargeant

Abstract

Autophagic flux, a crucial cellular process that is often overlooked in its significance for human health, has been linked to various chronic diseases, including cancer, heart disease, and dementia. Despite its importance, direct measurement in humans has been lacking, which has hindered translational efforts. Here, we describe a protocol for the measurement of human autophagic flux in blood. By treating whole blood with chloroquine and isolating peripheral blood mononuclear cells, autophagic flux can be measured by quantifying the autophagy protein LC3BII. This approach preserves individual genetic and nutritional parameters. This method can be used to identify factors influencing autophagic flux in humans, facilitating clinical translation and potentially serving as a biomarker for age-related chronic diseases. In this protocol, we describe detailed methodology, along with risks and limitations of this technique.

Key words Autophagy, Lysosome, Autophagic flux, Human, Blood, PBMC

1 Introduction

Autophagy is a pathway that sequesters material from the cytoplasm (such as damaged organelles and protein aggregates) and delivers it to the lysosome for hydrolysis. In this way, it slows biological aging [1]. The term "autophagy" was coined by Christian de Duve in 1963, who received the Nobel Prize in 1974 for his work on lysosomal biology [2]. In 2016, Yoshinori Ohsumi received the Nobel Prize for his work on understanding the machinery that drives the process of autophagy [3]. In the decades since de Duve coined the term, research on autophagy has increased exponentially as researchers have increasingly recognized its potential significance to human health and disease [4].

Autophagy is a challenging field of study and prone to misinterpretation due to the difficulty in its measurement. A protein that is commonly used to measure autophagy is microtubule-associated

Stephanie M. Hughes and Indranil Basak (eds.), *Lysosomes: Methods and Protocols*, Methods in Molecular Biology, vol. 2976, https://doi.org/10.1007/978-1-0716-4844-5_6,

protein 1 light chain 3 beta (MAP1LC3B/LC3B). During the process of autophagy, LC3B is proteolytically cleaved to form soluble LC3BI, which is then conjugated to phosphatidylethanolamine to form LC3BII [5], firmly anchoring it to the surface of the autophagosome, for which it is a robust marker [6]. As LC3BII is anchored to the luminal membrane and to the cytosolic membrane of the autophagosome, some of it is degraded when the lysosome fuses with the autophagosome. While LC3BII is a good marker for autophagosomes, its measurement does not equate to a measure of autophagy. In fact, in a seminal paper, Tanida and colleagues [7] demonstrated that increased autophagy could result in either increased or decreased steady-state amounts of LC3BII depending on the cell type. Instead, they recommended that autophagy should be measured by comparing how much LC3BII is accumulated in the presence of a lysosomal inhibitor. Only by looking at autophagic flux in this way could autophagy be reliably estimated. This technique has been used in cell culture systems and even in animals and has generated valuable preclinical data about autophagy in basic cell processes, aging, and disease.

While many thousands of research papers have been published on autophagy and our understanding of its basic mechanisms has enjoyed much progress in recent years, the translation of this preclinical knowledge to humans has been conspicuous by its absence. A major roadblock to its translation to humans has been the lack of a physiological human autophagy measure. Because autophagy is nutrient sensitive, measurement of primary human cells in artificial culture medium that often contains supraphysiological nutrient concentrations will not give a physiologically relevant measure of autophagy. We aimed to address this issue by optimizing the use of lysosomal inhibition in whole human blood for the measurement of autophagic flux in peripheral blood mononuclear cells (PBMCs) [8, 9]. As autophagy is sensitive to nutrient signaling [10], inhibiting lysosomal degradation to calculate autophagic flux in whole blood is important because it maintains a physiological environment. This protocol describes the measurement of autophagic flux in whole human blood. This involves treatment of whole blood with a lysosomal inhibitor, isolation of PBMC from the blood, cell lysis, protein quantification, and measurement of autophagy protein LC3BII with a commercially available ELISA kit. It is useful for human studies that seek to understand how autophagy correlates with factors such as aging and disease, or how it can be modified using medications or nutritional and other lifestyle interventions.

2 Materials

2.1 Lysosomal Inhibition in Whole Blood

1. Blood collection tubes: 9-mL vacuette lithium heparin blood collection tubes.
2. Chloroquine (CQ): water containing 50 mM CQ diphosphate salt. Store aliquoted at −20 °C.
3. Tube revolver rotator: Thermo Fisher Scientific.

2.2 PBMC Isolation

1. Dulbecco's Phosphate Buffered Saline (DPBS).
2. Lymphoprep (Stem Cell Technologies).
3. Red blood cell lysis buffer (1×): dilute BD Pharm Lyse Lysing buffer (10×) (BD Bioscience) to 1× in sterile water; pH between 7.1 and 7.4, adjust if necessary. Store excess 1× solution at 4 °C for up to 30 days.

2.3 Saponin Wash for Cell Pellets

1. PBS, pH 7.4: 137 mM NaCl, 2.7 mM KCl, 10 mM Na_2HPO_4, 1.8 mM KH_2PO_4.
2. PBS containing 0.05% (w/v) saponin: aliquot and store at −20 °C (*see* **Note 1**).
3. Protease inhibitors: cOmplete, EDTA-free Protease Inhibitor Cocktail Tablets (Roche). Dissolve 1 tablet in 2 mL of water to prepare a 25× stock solution. Aliquot and store at −80 °C.
4. Phosphatase inhibitors: 100 mM sodium orthovanadate (100× stock solution; aliquot and store at −20 °C), 100 mM sodium pyrophosphate (40× stock solution; store at 4 °C), and 1 M β-glycerophosphate, pH 7.4 (1000× stock solution; store at 4 °C).

2.4 Preparing Cell Lysates

1. 1× Cell Extraction Buffer: reagents included with the FastScan Total LC3B ELISA kit (Cell Signaling Technology): 7.4 mL of MilliQ water, 2 mL of FastScan ELISA Cell Extraction Buffer (5×), 200 μL of FastScan ELISA Cell Extraction Enhancer Solution (50×), and 400 μL of protease inhibitors (25×). Prepare fresh and allow to cool on ice before use. Adjust volumes as needed. Solution is stable for 1–2 weeks at 4 °C.
2. Protease inhibitors: cOmplete, EDTA-free Protease Inhibitor Cocktail Tablets (Roche). Dissolve 1 tablet in 2 mL of water to prepare a 25× stock solution. Aliquot and store at −80 °C.

2.5 Micro BCA Protein Assay to Determine Protein Concentration

1. Micro BCA protein assay kit (Thermo Fisher Scientific) containing: Micro BCA reagent A, Micro BCA reagent B, Micro BCA reagent C, 2 mg/mL bovine serum albumin standard (BSA) ampoules.

2. Reagent ABC: aliquot 5 mL of Micro BCA reagent A, 4.8 mL of Micro BCA reagent B, and 200 μL of Micro BCA reagent C into a 10-mL yellow-top tube. Vortex to mix. This will make 10 mL of color reagent, which is sufficient for one 96-well plate. Adjust volumes as required for the percentage of the plate used.

2.6 ELISA Measurement of LC3BII

1. FastScan Total LC3B ELISA Kit (Cell Signaling Technology, Cat # 35172).
2. FastScan ELISA Cell Extraction Buffer (5×): provided with the FastScan Total LC3B ELISA kit.
3. FastScan ELISA Cell Extraction Enhancer Solution (50×): provided with the FastScan Total LC3B ELISA kit.
4. Assay buffer: reagents provided with the FastScan Total LC3B ELISA kit: 7.8 mL of MilliQ water, 2 mL of FastScan ELISA Cell Extraction Buffer (5×), and 200 μL of FastScan ELISA Cell Extraction Enhancer Solution (50×). Make fresh just prior to starting the experiment. Calculate how much assay buffer is required before starting the experiment and adjust volumes accordingly.
5. Recombinant human LC3B (rhLC3B) standard: 0.5 mg/mL (batch-dependent concentration; Abcam). Store in 5-μL single-use aliquots at −80 °C.
6. Positive control (QC): reconstitute one vial of Positive Control (provided with the FastScan Total LC3B ELISA kit) in 250 μL of MilliQ water and store at −80 °C in 50-μL aliquots.
7. Capture antibody (4×): add 3 mL of FastScan ELISA Capture Antibody Diluent to the lyophilized capture antibody (both provided with the FastScan Total LC3B ELISA kit). Incubate at room temperature for 5 min. Gently mix.
8. HRP-linked antibody (4×): add 3 mL of FastScan ELISA HRP Antibody Diluent to the lyophilized HRP-linked antibody (both provided with the FastScan Total LC3B ELISA kit). Incubate at room temperature for 5 min. Gently mix.
9. Antibody cocktail: mix equal volumes of the 4× capture antibody and 4× HRP-linked antibody immediately prior to use. Reconstituted antibodies can be stored at 4 °C for up to 4 weeks; however, there may be a loss of signal.
10. Wash buffer: dilute 20× Wash Buffer (provided with the FastScan Total LC3B ELISA kit) to 1× with RO water.
11. TMB Substrate: provided with the FastScan Total LC3B ELISA kit.
12. Stop Solution: provided with the FastScan Total LC3B ELISA kit.

3 Methods

Refer to **Note 2** before commencing and adhere to all local ethics and safety guidelines for handling human blood and its derivatives, which are potentially infectious to laboratory workers.

3.1 Lysosomal Inhibition in Whole Blood

1. Collect subject blood (7–8 mL) in one 9-mL lithium heparin blood collection tube and mix gently by inverting the tube 5 times.
2. Maintain blood at room temperature in a vertical position until processing. Process within 1 h of blood collection.
3. Gently invert the blood tube 3 times and split the blood into two 3-mL aliquots by pipetting into 10-mL conical centrifuge tubes.
4. Treat one of the blood samples with 150 μM of CQ (9 μL of 50 mM CQ for 3 mL blood). Leave the other blood sample untreated. Mix gently by inverting both tubes 4 times.
5. Incubate both tubes (±CQ) at 37 °C for exactly 1 h with rotation on the tube revolver rotator (10 rpm).

3.2 PBMC Isolation

1. After the 1 h incubation, remove blood samples from the rotator and incubator (*see* **Note 3**) and keep on ice for the rest of the PBMC isolation.
2. Add 3 mL cold DPBS (ratio 1:1) into blood tubes and mix gently by inverting the tubes 4 times (*see* **Note 4**).
3. Prepare 4 mL of Lymphoprep in a 10-mL conical centrifuge tube.
4. Using a 5-mL pipette, carefully overlay the Lymphoprep with the blood mixture (*see* **Note 5**).
5. Repeat for both blood tubes.
6. Centrifuge for 30 min at 800× *g* at 4 °C, with brake off.
7. Using a 1-mL pipette, carefully aspirate the PBMC layer at the interface of the plasma and Lymphoprep and dispense in a 10 mL conical centrifuge tube (approximately 2 mL) (Fig. 1). Discard the remaining blood.
8. Dilute PBMC in cold DPBS to a final volume of 5 mL. Mix gently by inverting the tubes 4 times.
9. Centrifuge at 600× *g* for 10 min at 4 °C (with the brake on).
10. Aspirate and discard the supernatant fluid carefully.
11. Resuspend the pellet with 2 mL of 1× red blood cell lysis buffer and mix by gentle pipetting with a 1-mL pipette.
12. Incubate on ice for 5 min.
13. Add 5 mL of cold DPBS.

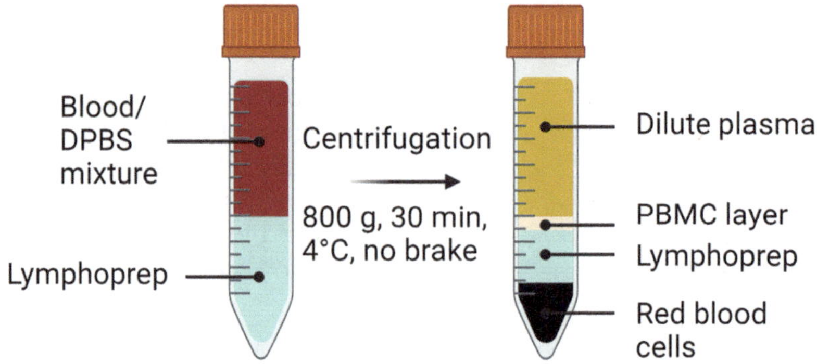

Fig. 1 Centrifugation of a blood/PBS mixture layered over the top of Lymphoprep yields the PBMC layer, which is used in downstream steps. Created with BioRender.com

14. Centrifuge at 600× *g* for 5 min at 4 °C.
15. Aspirate and carefully discard the supernatant fluid.
16. Resuspend the cell pellet with 1 mL of cold DPBS and mix gently using a 1-mL pipette.
17. Transfer the cells in DPBS into Eppendorf tubes.
18. Centrifuge at 2000× *g* for 10 min at 4 °C.
19. Using a 1-mL pipette, carefully aspirate the supernatant fluid.
20. Place the tubes containing the PBMC pellets on dry ice to rapidly freeze the samples.
21. Store PBMC pellets at −80 °C until analysis.

3.3 Saponin Wash for Cell Pellets

If the saponin-soluble fraction obtained in **step 4** is required for analysis of cytosolic components, please *see* **Note 6**.

1. Thaw PBMC pellets obtained in Subheading 3.2 on ice.
2. Resuspend cell pellets in 100 μL (room temperature) of PBS containing 0.05% saponin (with or without protease and phosphatase inhibitors, as per **Note 6**).
3. Incubate samples on a platform rocker for 5 min at room temperature.
4. Centrifuge at 16,000× *g* for 5 min at 4 °C. Remove and retain the saponin-soluble supernatant for analysis of cytosolic components, if required (*see* **Note 6**). Otherwise discard.
5. Resuspend the pellet in 100 μL of PBS.
6. Centrifuge at 16,000× *g* for 5 min at 4 °C. Discard supernatant.
7. Use saponin-washed cell pellets immediately for preparing cell lysates.

3.4 Preparing Cell Lysates

1. Resuspend the pellet in 100 μL of ice-cold 1× Cell Extraction Buffer.
2. Sonicate samples on ice for 20 sec, chill on ice, and then repeat sonication with a probe sonicator (*see* **Note 7**).
3. Centrifuge at 16,000× *g* for 5 min at 4 °C.
4. Retain supernatant—this is the final cell lysate. Discard pellet.

3.5 Micro BCA Protein Assay to Determine Protein Concentration

1. Dilute samples if required: to make dilutions, invert samples twice and vortex for 5 s prior to diluting. Use Milli-Q water as the diluent if required (*see* **Note 8**).
2. Make the BSA working standard by mixing 100 μL of 2 mg/mL BSA standard stock from the Micro BCA Protein Assay kit with 900 μL of Milli-Q water (200 μg/mL final concentration). Invert and vortex thoroughly.
3. Dilute the BSA working standard in Milli-Q water to give a standard curve (in duplicate) in rows A–H and columns 1 and 2 of a 96-well plate (*see* Table 1).
4. Add samples (which may require dilution, *see* **Note 8**) into the 96-well plate in duplicate.
5. To each well containing a sample, make the total well volume up to 100 μL with Milli-Q water.
6. Using a multichannel pipette, aliquot 100 μL of Reagent ABC into each well.
7. Seal plate with sealing film or sticky tape, ensuring the adhesive back is securely sealed over all 96 wells.
8. Incubate at 37 °C for 2 h or at 60 °C for 30 min.

Table 1
Volumes for calibration curve components for the Micro BCA protein assay procedure

	μg BSA/well	Volume of 200 μg/mL BSA stock added to the well	Milli-Q water (μL)	Total volume/well (μL)
A	0 μg	0 μL	100 μL	100 μL
B	1 μg	5 μL	95 μL	100 μL
C	2 μg	10 μL	90 μL	100 μL
D	4 μg	20 μL	80 μL	100 μL
E	8 μg	40 μL	60 μL	100 μL
F	12 μg	60 μL	40 μL	100 μL
G	16 μg	80 μL	20 μL	100 μL
H	20 μg	100 μL	0 μL	100 μL

9. Read the plate on a plate reader that measures absorbance at 560 nm.
10. Use the absorbance results from the calibration curve and samples to calculate protein concentration in cell lysates. For rejection criteria, *see* **Note 9**.

3.6 ELISA Measurement of LC3BII

1. Warm all FastScan Total LC3B ELISA kit reagents to room temperature before use.
2. Using the total protein concentration calculated in Subheading 3.5, dilute PBMC lysates down using Assay Buffer to 0.1 mg of protein/mL in a final volume of 200 μL. This is sufficient for three 50-μL replicates.
3. Prepare the recombinant human LC3 protein (rhLC3) standard as follows:
   - 3.1. Remove a 5-μL aliquot of 0.5 mg/mL (batch dependent) rhLC3 stock from the −80 °C freezer (*see* **Note 10**).
   - 3.2. Dilute the standard 1:10 by adding 2 μL of standard to 18 μL of assay buffer (*see* **Note 11**).
   - 3.3. Further dilute the standard 1:500 by adding 2 μL of the 1:10 diluted standard prepared in **step 3.2** to 998 μL of Assay Buffer. This will yield a final dilution of 1:5000, or 0.1 ng/μL.
4. Load the standard onto the ELISA plate (supplied with the FastScan Total LC3B ELISA kit) in triplicate to a final volume of 50 μL, as per Table 2.
5. Load samples (50 μL/well) onto the ELISA plate in triplicate. Vortex samples before loading and between each triplicate. This step is critical for minimizing variation within the triplicate.

Table 2
Volumes for calibration curve components for the LC3BII ELISA

ng of rhLC3/well	μL of standard from Subheading 3.6, step 3.3	μL of assay buffer
0	0	50
0.25	2.5	47.5
0.5	5	45
1.0	10	40
1.5	15	35
2.0	20	30
3.0	30	20
4.0	40	10

6. Load 5 μL of the positive control QC sample in triplicate onto the ELISA plate. Add 45 μL of Assay Buffer to make up to a final volume of 50 μL in the well.
7. Using a multichannel pipette, add 50 μL of the antibody cocktail mix to each well.
8. Cover plate with sealing tape (provided in the FastScan Total LC3B ELISA kit) and incubate on an orbital shaker at 150 rpm for 1 h at room temperature.
9. Decant or aspirate the contents from all wells. Wash the wells 4× with 250 μL 1× wash buffer, discarding the buffer between washes (*see* **Note 12**). Blot the plate on a paper towel after the last wash to remove excess liquid.
10. Using a multichannel pipette, add 100 μL of TMB substrate to each well.
11. Cover plate with sealing tape, wrap in aluminum foil, and incubate on an orbital shaker at 150 rpm for 15 min at room temperature.
12. Using a multichannel pipette, add 100 μL Stop Solution to each well. The solution will change from blue to yellow.
13. Read absorbance at 450 nm on a plate reader.
14. Results can be calculated in Excel or PRISM based on comparing values from samples to values from the standard curve. For rejection criteria, *see* **Note 13**.
15. Autophagic flux (the amount of LC3BII in ng turned over per mg of protein per hour) is calculated according to the following equation:

$$\Delta \text{LC3BII (ng/mg protein/h)} \\ = \text{LC3BII}_{+\text{CQ}}\text{(ng/mg protein)} - \text{LC3BII}_{-\text{CQ}}\text{(ng/mg protein)}$$

The time unit for ΔLC3BII is based on the accumulation of LC3BII during the 1-h incubation of whole blood with CQ, which inhibits the breakdown of LC3BII.

4 Notes

1. Weigh the saponin inside a fume hood: the dust is highly irritating to the respiratory tract.
2. Ensure the incubator is turned on and at 37 °C. Place the tube rotator inside the incubator. Thaw CQ at 37 °C in the incubator. Take DPBS from 4 °C and place on ice. Ensure there is sufficient 1× red blood cell lysis buffer; if not, prepare more (*see* Subheading 2.2, **item 3**). Collect dry ice and keep it ready in an

appropriate container. Gently mix Lymphoprep by inverting the bottle before use. Set centrifuges to 4 °C.

3. Restart the rotator if there are tubes remaining on the rotator for subsequent processing.
4. For quicker processing times and larger yields, the use of SepMate tubes (SepMate-15 (IVD), Stem Cell Technologies) is recommended for the PBMC isolation. In this case, instead of following Subheading 3.2, **steps 3**–7, use the method below:
   4.1. Place 4 mL of Lymphoprep in a SepMate tube by carefully pipetting it through the central hole of the SepMate insert. The top of the Lymphoprep will be above the insert.
   4.2. Keeping the SepMate tube vertical, add the diluted blood sample by slowly pipetting it down the side of the tube. The sample will sit above the insert.
   4.3. Repeat for both blood tubes.
   4.4. Centrifuge for 10 min at 1200× g at 4 °C, with the brake on.
   4.5. Pipette off some of the plasma above the PBMC layer to reduce contamination and discard.
   4.6. Pour off the top layer containing the PBMC into a 10-mL conical centrifuge tube. Do not hold the SepMate tube inverted for longer than 2 s. Discard the remaining blood.
   4.7. Continue from Subheading 3.2, **step 8**.
5. Overlaying the blood mixture with Lymphoprep must be done slowly and carefully to minimize mixing between the two phases.
6. The purpose of the saponin wash is to remove soluble proteins contained in the cytosol of the cells. This includes LC3BI. What remains in the cell pellet after the saponin wash contains membrane-bound proteins, including LC3BII bound to autophagosomes. Bensalem et al. [9] demonstrate partitioning of LC3BI from LC3BII using this method. It may be necessary to retain the saponin wash fraction obtained in Subheading 3.3, **step 4** for future analysis, for example, western blotting for cytosolic proteins. If so, the saponin used in Subheading 3.3, **step 2** should also have protease inhibitors and phosphatase inhibitors added to it before use. During Subheading 3.3, **step 4**, transfer the saponin-soluble supernatant to a new Eppendorf tube and process according to requirements.
7. The probe sonicator is extremely noisy: ear protection is highly recommended. Sonication should be done in a biosafety hood to reduce the risk of aerosol inhalation.

8. Normally, no dilution is required. Assay 5 μL (neat) of the cell lysate.
9. Rejection criteria for Micro BCA protein concentration assay. Reject a sample if the:
   - % CV value of the replicates is >10%.
   - concentration value is NOT between 2 and 16 μg/volume sampled.
10. The rhLC3 standard aliquots are single use. Do not re-freeze as LC3 is unstable during freeze-thaw cycles.
11. The stock rhLC3 standard concentration is batch-dependent. Therefore, adjust the dilution accordingly to give a diluted concentration of 0.05 mg/mL at this step.
12. An automated plate washer can be used for the wash steps, if available.
13. Rejection criteria for LC3B ELISA:
    - Before unblinding data, highlight data with a %CV value higher than 10% and remove the most divergent of the triplicates. If, after this, % CV value of the duplicate is >15%, reject the sample.
    - Reject the sample if the concentration value is outside of the standard curve.

Conflict of Interest

TJS and JB are listed as inventors on a related patent for this method—Australia (Provisional) 2019903187; 2019904822; PCT/AU/2020/050908; United Kingdom GB2204321.0; USA 17/637,494.

References

1. López-Otín C, Blasco MA, Partridge L, Serrano M, Kroemer G (2023) Hallmarks of aging: an expanding universe. Cell 186(2): 243–278. https://doi.org/10.1016/j.cell.2022.11.001
2. Klionsky DJ (2008) Autophagy revisited: a conversation with Christian de Duve. Autophagy 4(6):740–743. https://doi.org/10.4161/auto.6398
3. Levine B, Klionsky DJ (2017) Autophagy wins the 2016 Nobel prize in physiology or medicine: breakthroughs in baker's yeast fuel advances in biomedical research. Proc Natl Acad Sci USA 114(2):201–205. https://doi.org/10.1073/pnas.1619876114
4. Klionsky DJ, Petroni G, Amaravadi RK, Baehrecke EH, Ballabio A, Boya P, Bravo-San Pedro JM, Cadwell K, Cecconi F, Choi AMK, Choi ME, Chu CT, Codogno P, Colombo MI, Cuervo AM, Deretic V, Dikic I, Elazar Z, Eskelinen EL, Fimia GM, Gewirtz DA, Green DR, Hansen M, Jäättelä M, Johansen T, Juhász G, Karantza V, Kraft C, Kroemer G, Ktistakis NT, Kumar S, Lopez-Otin C, Macleod KF, Madeo F, Martinez J, Meléndez A, Mizushima N, Münz C, Penninger JM, Perera RM, Piacentini M, Reggiori F, Rubinsztein DC, Ryan KM, Sadoshima J, Santambrogio L, Scorrano L, Simon HU, Simon AK, Simonsen A, Stolz A, Tavernarakis N, Tooze SA, Yoshimori T,

Yuan J, Yue Z, Zhong Q, Galluzzi L, Pietrocola F (2021) Autophagy in major human diseases. The EMBO journal:e108863. https://doi.org/10.15252/embj.2021108863

5. Ichimura Y, Kirisako T, Takao T, Satomi Y, Shimonishi Y, Ishihara N, Mizushima N, Tanida I, Kominami E, Ohsumi M, Noda T, Ohsumi Y (2000) A ubiquitin-like system mediates protein lipidation. Nature 408(6811): 488–492. https://doi.org/10.1038/35044114
6. Kabeya Y, Mizushima N, Ueno T, Yamamoto A, Kirisako T, Noda T, Kominami E, Ohsumi Y, Yoshimori T (2000) LC3, a mammalian homologue of yeast Apg8p, is localized in autophagosome membranes after processing. EMBO J 19(21):5720–5728. https://doi.org/10.1093/emboj/19.21.5720
7. Tanida I, Minematsu-Ikeguchi N, Ueno T, Kominami E (2005) Lysosomal turnover, but not a cellular level, of endogenous LC3 is a marker for autophagy. Autophagy 1(2): 84–91. https://doi.org/10.4161/auto.1.2.1697
8. Bensalem J, Hattersley KJ, Hein LK, Teong XT, Carosi JM, Hassiotis S, Grose RH, Fourrier C, Heilbronn LK, Sargeant TJ (2021) Measurement of autophagic flux in humans: an optimized method for blood samples. Autophagy 17(10):3238–3255. https://doi.org/10.1080/15548627.2020.1846302
9. Bensalem J, Teong XT, Hattersley KJ, Hein LK, Fourrier C, Liu K, Hutchison AT, Heilbronn LK, Sargeant TJ (2023) Basal autophagic flux measured in blood correlates positively with age in adults at increased risk of type 2 diabetes. Geroscience 45. https://doi.org/10.1007/s11357-023-00884-5
10. Hein LK, Apaja PM, Hattersley K, Grose RH, Xie J, Proud CG, Sargeant TJ (2017) A novel fluorescent probe reveals starvation controls the commitment of amyloid precursor protein to the lysosome. Biochim Biophys Acta, Mol Cell Res 1864(10):1554–1565. https://doi.org/10.1016/j.bbamcr.2017.06.011

Chapter 7

Lysosome Enrichment Using Superparamagnetic Iron Oxide Nanoparticles (SPIONs)

Pathma Muthukottiappan, Sara Bonini, and Dominic Winter

Abstract

Lysosomes are dynamic and low-abundant organelles, and their enrichment is essential to facilitate a comprehensive biochemical characterization. Several methods are available to achieve this goal, such as density gradient centrifugation, immunoprecipitation, and superparamagnetic iron oxide nanoparticles (SPIONs). Compared to the other approaches, SPIONs present several advantages, as they enable the fast enrichment of intact lysosomes at high yields and purity without the need for the expression of tagged proteins. This protocol describes an optimized workflow for the enrichment of lysosomes from mammalian cells using SPIONs. We describe cell culture and treatment conditions tailored toward commonly utilized cell lines, approaches for cell rupture and lysosome enrichment, and downstream analyses by enzymatic assays and western blotting to assess the success of lysosome enrichment.

Key words SPIONs, Lysosomes, Endocytosis, Superparamagnetic nanoparticles, Dextran, β-hexosaminidase assay

1 Introduction

Lysosomes are dynamic and low-abundant organelles. Based on immunostaining and mass spectrometry-based proteomics analyses, respectively, it was estimated that a given cell contains a few hundred lysosomes [1], which comprise ~0.2% of cellular protein mass [2]. Consequently, individual lysosomal proteins are typically low-abundant. The lysosomal proteome furthermore presents with a high dynamic range, both within the same cell (when different lysosomal proteins are compared to each other) and between cell lines (when levels of the same protein are compared), complicating its analysis [3]. Therefore, to enable the comprehensive biochemical characterization of the lysosomal proteome, or the lysosome's metabolic content and features, organelle enrichment is a prerequisite. In this context, three major strategies are frequently used: density gradient centrifugation, immunoprecipitation through

Stephanie M. Hughes and Indranil Basak (eds.), *Lysosomes: Methods and Protocols*, Methods in Molecular Biology, vol. 2976, https://doi.org/10.1007/978-1-0716-4844-5_7,

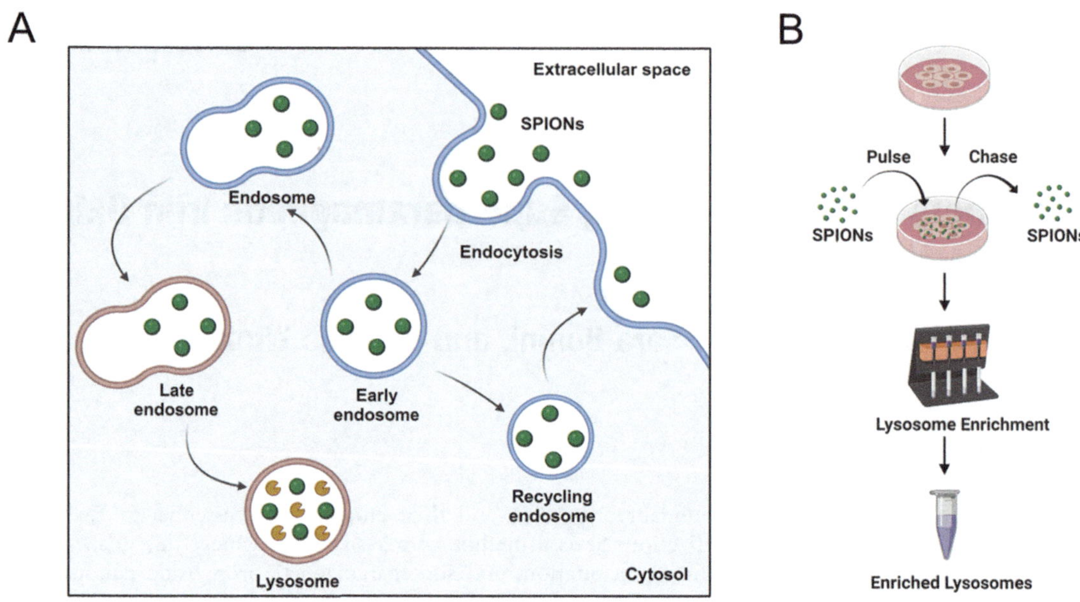

Fig. 1 SPION-mediated enrichment of lysosomes. (**a**) Delivery of SPIONs to lysosomes. (**a**) SPIONs are internalized via unspecific fluid-phase endocytosis and are initially present in early endosomes. They are then transported to late endosomes, which eventually fuse with lysosomes, making the latter the ultimate destination for SPIONs. (**b**) Workflow for the enrichment of lysosomes by SPIONs. After delivery of SPIONs to lysosomes, the plasma membrane is selectively ruptured, and lysosomes are enriched using a magnetic field. (Generated with Biorender.com)

tagged membrane proteins, and superparamagnetic iron oxide nanoparticles (SPIONs) [4].

Compared to the other mentioned approaches, SPIONs come with several advantages. They do not require any expression of tagged proteins and enable the enrichment of intact lysosomes for follow-up studies. Furthermore, they provide a fast enrichment procedure with a high yield [4]. For lysosome enrichment, SPIONs are added to the cell culture medium, from which they are taken up by cells through the endocytic pathway, and are ultimately delivered to lysosomes (Fig. 1a) [5]. In this context, lysosomal enrichment efficiency largely relies on endocytosis rates and endolysosomal transport kinetics. Hence, by adjustment of pulse chase conditions, it is possible to achieve near-exclusive presence of SPIONs in lysosomes, enabling the efficient enrichment of terminal lysosomes [6] (Fig. 1b). On the other hand, this renders SPIONs incompatible with cell culture models that have a dysfunctional endocytic machinery.

SPIONs have been employed since 1998 in numerous studies to enrich and characterize lysosomes (reviewed by [7]). They are typically <40 nm in size and consist of an iron oxide core that can be reversibly attracted using an external magnetic field, enabling their efficient enrichment through the application of magnetic force

[5, 8]. An important feature of SPIONs is their surface coating, which typically consists of biocompatible substances like dextran or meso-2,3-dimercaptosuccinic acid (DMSA) [5, 9]. These coatings are crucial to prevent hydrophobic interactions, which can lead to the aggregation of SPIONs [10], and to facilitate their stability, attachment to the cell surface, and endocytosis. Further crucial features of SPIONs are their size, which is directly negatively proportional to their magnetism, and their oxidation state, as oxidation of their main constituent magnetite (Fe_3CO_4) to maghemite (γ-Fe_2CO_3) reduces magnetic strength. Finally, uncoated SPIONs were shown to be degraded in the lysosomal lumen, resulting in the release of free $Fe^{2+/3+}$, which was speculated to potentially trigger the formation of reactive oxygen species [11]. However, a recent investigation of the effects of dextran-coated SPIONs on the lysosomal proteome and its oxidation status did not identify any detrimental effects [16].

2 Materials

Prepare all solutions in double-distilled water (ddH_2O) at room temperature (RT) unless indicated otherwise.

2.1 Cell Culture and Treatment of SPIONs

1. Human embryonic kidney (HEK) 293 cells.
2. Human epithelial cells (HeLa).
3. Mouse embryonic fibroblasts (MEF).
4. Mouse embryonic fibroblasts from a mouse NIH/Swiss embryo (NIH/3 T3).
5. Human epithelial-like hepatoblastoma cells (HuH-7).
6. Human neuroblastoma cells (SH-SY5Y).
7. 1× phosphate-buffered saline (PBS), pH 7.4.
8. Poly-L-lysine.
9. Dulbecco's modified Eagle medium (DMEM).
10. Fetal calf serum (FCS).
11. 10,000 IU/mL penicillin, 10 mg/mL streptomycin (100×).
12. 200 mM L-glutamine (100×).
13. 0.05% Trypsin–EDTA (1×).
14. DexoMAG 40 (SPIONs, Catalog no: BKS25, Liquids Research LTD).

2.2 Harvesting, Dounce Homogenization of Cells, and Lysosome Enrichment

1. Isolation Buffer A: 250 mM sucrose, 10 mM HEPES (pH 7.5), 15 mM $CaCl_2$, 1.5 mM MgAc, 1 mM $MgCl_2$, 15 mM KCl, 1× cOmplete, EDTA-free protease inhibitor cocktail, 1× phosphatase inhibitor cocktail (PhosSTOP), 10 mM β-glycerophosphate, 10 mM sodium orthovanadate, and 10 mM pyrophosphate (*see* **Note 1**).

2. Isolation Buffer B: 20 mM HEPES (pH 7.5), 120 mM KCl, 10 mM NaCl, 2 mM MgCl, 1× cOmplete, EDTA-free protease inhibitor cocktail, 1× PhosSTOP, 10 mM β-glycerophosphate, 10 mM sodium orthovanadate, and 10 mM pyrophosphate (*see* **Note 1**).
3. 15 mL of dounce homogenizer with a tight pestle.
4. LS columns (Catalog no: 130-042-401, Miltenyi Biotech).
5. MACS MultiStand (Catalog no: 130-042-303, Miltenyi Biotech).
6. 0.5% bovine serum albumin (BSA) in 1× PBS.
7. Refrigerated centrifuge, swinging bucket rotor SX4750 for 15-mL tubes.
8. Refrigerated centrifuge, FA-24x2 rotor for 1.5-mL tubes.

2.3 β-Hexosaminidase Assays and Western Blotting

1. Reaction Solution: 100 mM sodium citrate, 0.2% BSA, 10 mM para-nitrophenyl-N-acetyl-2-B-D-glucosaminide, pH 4.6.
2. 0.1% Triton X-100 in ddH_2O.
3. Stop Solution: 0.4 M glycine, pH 10.4.
4. 10% sodium dodecyl sulfate–polyacrylamide precast gel.
5. 1× SDS running gel buffer: 25 mM Tris–HCl, 192 mM glycine, 0.1% SDS, pH 8.3.
6. Mini-Protean Tetra Vertical Electrophoresis Cell.
7. Blotting membranes, Amersham Hybond 0.2 μm PVDF.
8. Amersham Protran Western blotting membranes, 0.2-μm nitrocellulose.
9. ROTILABO Blotting papers, 0.35 mm.
10. Wet transfer buffer: 25 mM Tris, 192 mM glycine, 20% methanol.
11. Semi-dry transfer buffer: 48 mM Tris base, 39 mM glycine, 20% methanol.
12. Trans-Blot Turbo Transfer System.
13. Wet/Tank Blotting System.
14. Clarity Western ECL Substrate.
15. Antibodies: ATP6V1B2 (Catalog no. sc-166045, Santa Cruz, 1:1000), ATP6V1D (Catalog no. 14920-1-AP, Proteintech, 1: 1000), Calnexin (Catalog no. 66903-1-AP, Proteintech, 1: 20,000), CTSD (Catalog no. ab97499, Abcam, 1:1000), GBA (Catalog no. 27972-1-AP, Proteintech, 1:1000), GM130 (Catalog no. 610822, BD, 1:1000), LIMP2 (Catalog No. AF1966, R&D, 1:400), NPC2 (Catalog No. 19888-1-AP, Proteintech, 1:500), and SDHA (Catalog no. 14865-1-AP, Proteintech, 1:800).

16. 96-well plate.
17. Spectra FLUOR plus microplate reader.

3 Methods

3.1 Cell Seeding and Labeling of SPIONs

Cell density is a crucial factor for the success of the experiment. If it is too low or high, a reduced yield of enriched lysosomes can be observed, which is most likely due to undesired rupture of cells during harvesting or limited surface area for interaction with SPIONs during cell culture, respectively. Based on the type of cell line, different cell numbers and timepoints of seeding (with respect to the SPIONs treatment) are optimal (Table 1). Furthermore, some cell lines require coating of cell culture dishes to increase cell adhesion and to prevent cells from detaching due to the repetitive medium exchange steps during treatment of SPIONs.

1. Dissolve PLL in 1× PBS to a final concentration of 5 mg/mL and filter-sterilize using a 0.2-μm filter under a sterile hood.
2. Dilute the PLL solution using 1× PBS to a final concentration of 0.5 mg/mL (*see* **Note 2**).
3. Add 3–4 mL of PLL solution to each 10 cm cell culture dish and incubate for 20 min at 37 °C.
4. Aspirate the PLL solution and wash three times with 1× PBS (*see* **Note 3**).

Table 1
Cell culture conditions for different cell lines for enriching lysosomes using SPIONs. For all cell lines, a SPION pulse/chase duration of 24 h is applied

Cell line	Number of cells seeded for each 10-cm dish	Doubling time, h	% FCS after cell seeding (before SPION pulse)	SPION addition (h after cell seeding)
HEK393	6×10^6	24	10	0
HeLa	3×10^6	32	10	24
MEF	3×10^6	48	2.5	72
NIH/3 T3	6×10^6	24	2.5	72
HuH-7	5×10^6	24	10	24
SH-SY5Y	6×10^6	48	10	24

Adapted from Ref. [3]

5. Prepare complete medium by combining DMEM with 10% FCS, 100 IU/mL penicillin, 100 μg/mL streptomycin, and 2 mM L-glutamine (final concentrations).
6. Seed 6×10^6 HEK293 cells per 10-cm dish using complete medium containing SPIONs (final concentration: 10%) (for cell numbers of other cell lines, see Table 1). To obtain optimal results, seed four 10-cm dishes for one enrichment experiment (*see* **Note 4**).
7. Incubate cells in complete medium containing SPIONs for 24 h (for durations of other cell lines, *see* Table 1) (*see* **Note 5**).
8. Carefully aspirate the SPION-containing medium and wash three times with 1× PBS (*see* **Note 6**).
9. Supplement cells with fresh complete medium and incubate for 24 h (for durations of other cell lines, *see* Table 1) (*see* **Note 7**).
10. Carefully aspirate the complete medium, place the cell culture dish on ice, and wash cells once with 10 mL of ice-cold 1× PBS, and aspirate as much liquid as possible.
11. Add 2 mL of Isolation Buffer A to each 10 cm dish. Use a cell scraper to detach cells and pool the cell suspensions obtained from four 10-cm dishes in a 15-ml tube (*see* **Note 8**).

3.2 Homogenization of Cells and Lysosome Enrichment

It is of crucial importance to rupture the plasma membrane while keeping lysosomes intact. Therefore, the cell homogenization technique and the choice of buffers significantly impact the efficiency of lysosomal enrichment. Typically, isotonic or hypotonic isolation buffers containing sucrose are used for cell homogenization, dependent on the stability of the individual cells' membranes and if swelling of cells is desired (which can assist in disrupting the plasma membrane) [12, 13].

Keep the samples and all solutions on ice during the homogenization and enrichment process, except the LS columns and the magnetic stand.

1. Transfer the pooled cell suspension (8 mL) to a 15-mL dounce homogenizer and homogenize cells with 25 strokes (*see* **Note 9**). Transfer 200 μL of this fraction (total fraction) to a microtube and store on ice for quality control (QC) experiments (do not freeze the sample, *see* **Note 10**). Transfer the complete sample to a 15-mL tube and centrifuge at 4 °C, 500× *g* for 10 min (*see* **Note 11**). Collect the post-nuclear supernatant (PNS) and transfer it to a fresh tube.
2. Resuspend the pellet containing intact cells in 4 mL of Isolation Buffer A, transfer to the 15-mL dounce homogenizer, and homogenize for 25 strokes. Centrifuge at 4 °C, 500× *g* for 10 min. Pool the supernatant (12 mL) with the PNS from **step 1**. Transfer 200 μL of the PNS to a microtube and store on ice for QC (*see* **Notes 10** and **12**).

3. Place LS columns into the MACS MultiStand and equilibrate the columns using 1 mL of 0.5% BSA each at room temperature (RT) by gravity flow.
4. Place a 15-mL tube under the column, gently pour the PNS onto the column, and allow it to pass by gravity flow.
5. Collect the flow through (FT) fraction and reapply it to the same column (*see* **Note 13**). Collect the FT in a 15-mL tube, transfer 200 μL to a microtube, and store on ice for QC (*see* **Note 10**).
6. Wash the column with 1 mL of fresh Isolation Buffer A three times. After adding Isolation Buffer A, allow it to pass through the column by gravity. Collect the first 1 mL of wash fraction in a microtube and store on ice for QC (*see* **Note 10**).
7. Remove the column from the MACS MultiStand and tap it against a hard surface (*see* **Note 14**). Add 500 μL of Isolation Buffer A to the column and press it through using a plunger (takes typically 5–10 s) to elute lysosomes into a microtube placed on ice. Repeat this step and pool both eluate fractions (resulting in a total eluate volume of 1 mL). Transfer 200 μL of the eluate fraction to a microtube and store on ice for QC (*see* **Note 10**).
8. Dependent on the subsequent experiments, enriched lysosomes can be pelleted by centrifugation at 20,000× *g* for 30 min at 4 °C *(optional, see* **Note 15**).

3.3 Assessment of Lysosomal Recovery and Intactness by Enzymatic β-Hexosaminidase Assays

β-Hexosaminidase is a lysosomal luminal hydrolase, which catalyzes the hydrolysis of nonreducing ends of N-acetylglucosamine and N-acetylgalactosamine residues of different glycoconjugates [14]. Rupture of the lysosomal membrane during cell homogenization results in the release of β-hexosaminidase. The lysosomal membrane can also be solubilized by the addition of detergents, such as Triton X-100, to the sample, also resulting in the release of luminal hydrolases. Consequently, the difference in enzymatic β-hexosaminidase activity of samples with/without the addition of Triton X-100 can be used to assess the fraction of broken/intact lysosomes, respectively, in individual fractions. β-Hexosaminidase activity can be determined by an enzymatic assay utilizing Para-Nitrophenyl-N-acetyl-2-B-D-Glucosaminide, which releases p-nitrophenol, which can be detected at 405 nm upon ionization in basic pH [15]. Based on individual readings, the enzymatic activity can be calculated using Eq. 1.

$$\text{Intact ratio, \%} = 100 - \left(\left(\frac{\text{mU of eluate fraction without Triton}}{\text{mU of eluate fraction with Triton}}\right) * 100\right)$$

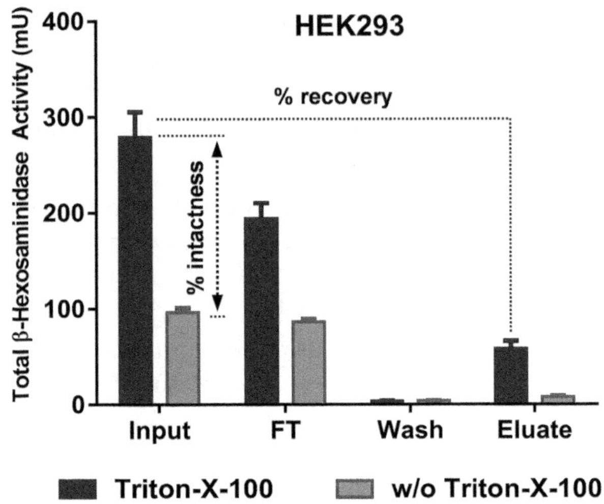

Fig. 2 Quality control of lysosome enrichment by β-hexosaminidase assay. The activity of the lysosomal luminal enzyme β-hexosaminidase is assessed in fractions obtained during the enrichment process. The difference in the enzymatic activity between samples with/without the addition of Triton-X 100 (disrupted/intact) enables the determination of the amount of intact lysosomes. (Adapted from Ref. [3])

$$\beta - \text{hexosaminidase activity, mU} = \left(\frac{(\text{AVR} * \text{CF} * \text{V})}{(\varepsilon * \text{VF} * \text{T})}\right) * \text{TF}$$

Equation 1 *Calculation of β-hexosaminidase activity.* β-Hexosaminidase activity (mU) is calculated using Beer Lambert's law using the following parameters: AVR: average of readings; CF: conversion factor (1,000,000); V: total reaction volume (μL) in each well; ε: molar extinction coefficient of the substrate (18,500); VF: volume (μL) of the sample (typically 25 μL); T: incubation time in min (typically 15 min); TF: total volume of each fraction obtained during the enrichment process (mL).

An example of the β-hexosaminidase assay readout is shown in Fig. 2. All solutions must be kept on ice during pipetting.

1. Add 25 μL of the QC fractions collected in Subheading 3.2 in four separate wells of a transparent 96-well plate (*see* **Note 16**).
2. Add 8 μL of 0.1% Triton X-100 to two wells.
3. Add 50 μL of Reaction Solution to each well and incubate for 15 min at 37 °C.
4. Add 200 μL of Stop Solution to each well.
5. Read the absorbance using the Spectra FLUOR plus microplate reader with the following measurement parameters: Absorbance: 405 nm; Gain: 60; Number of flashes: 3.

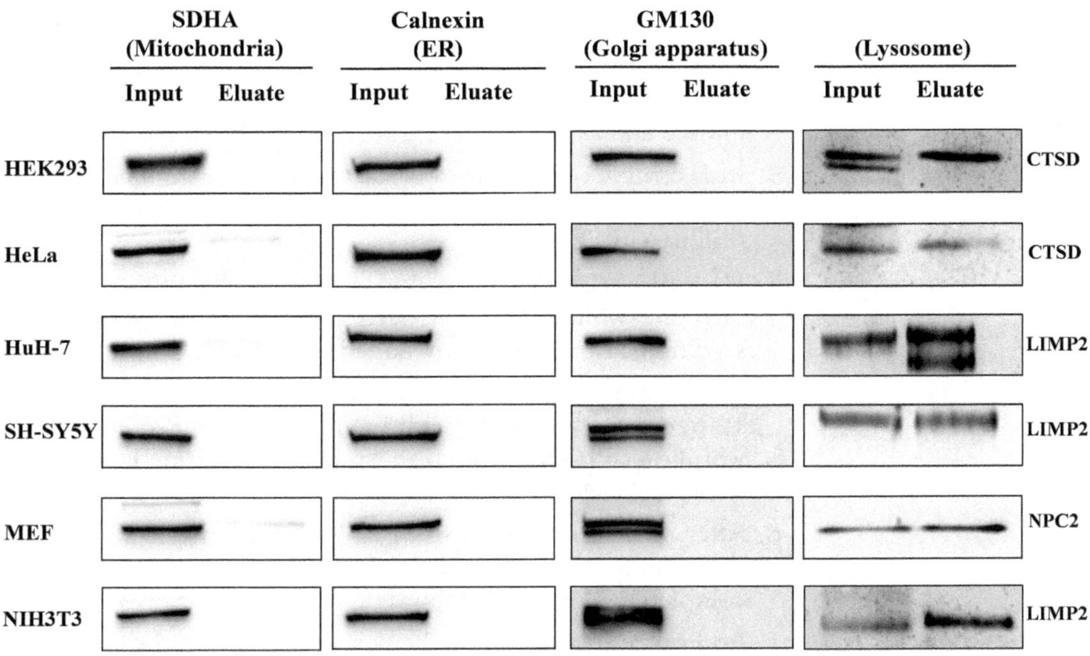

Fig. 3 Verification of lysosome enrichment by western blot analysis. Western blot analysis of post-nuclear supernatant (input) and eluate (enriched lysosomes) fractions probed with antibodies for organelle-specific marker proteins. Based on the individual protein abundances, enrichment of lysosomes and depletion of mitochondria, endoplasmic reticulum (ER), and Golgi apparatus can be observed. For each fraction, 1% of the respective total volume was subjected to western blot analysis. To further assess the recovery of intact lysosomes, antibodies against the lysosomal luminal proteins (CTSD and NPC2) and the lysosomal membrane protein 2 (LIMP2) were used. (Adapted from Ref. [3])

3.4 Assessment of Lysosome Enrichment Efficiency and Contamination with Other Organelles by Western Blot Analysis

In order to evaluate lysosomal purity and enrichment, the presence of marker proteins for contaminating other organelles or lysosomes, respectively, can be assessed by western blotting. Depending on the cell type and density, ~0.5–2% of each QC fraction obtained during the enrichment process should be used for SDS-PAGE and western blotting. Depending on the antibody, either wet or semi-dry blotting delivers optimal results (for details, *see* [3]). Successful lysosome enrichment should result in an increase and a decrease of signal intensities for markers of lysosomes and other organelles, respectively (Fig. 3).

4 Notes

1. Protease/phosphatase inhibitors are not stable for prolonged times at RT or 4 °C in solution. It is, therefore, recommended to add them to the isolation buffers immediately before cell lysis. Phosphatase inhibitors such as PhosSTOP, β-glycerophosphate, sodium orthovanadate, and

pyrophosphate are only necessary if you are interested in studying phosphoproteins.

2. The diluted PLL solution can be used up to 5 times; both unused and already used solutions should be stored at −20 °C.
3. PLL-coated cell culture dishes can be used immediately or stored at 4 °C for up to 2 weeks by addition of 1× PBS and sealing with Parafilm.
4. HEK293 cells can be seeded directly in SPION-containing medium, while other cells need a certain recovery time after seeding before SPIONs are added (Table 1).
5. Incubation of cells in SPION-containing medium for less than 12 h reduces lysosomal yield. Incubation times exceeding 24 h will allow SPIONs to reach all lysosomes, resulting in increased lysosomal yield.
6. After incubating cells with SPIONs, make sure to wash them thoroughly (but carefully) three times with 1× PBS to completely remove SPIONs.
7. Additional treatments can be performed during the chase time after the addition of SPIONs. In case biological treatments are not compatible with the cell numbers indicated in Table 1, the number of cell culture dishes used as input for one SPION enrichment experiment should be adjusted accordingly. It is recommended to use cells from four to eight 10-cm dishes per enrichment.
8. Lysosomal yield depends on the number of cells utilized for homogenizing and the ratio of cell pellet to Isolation Buffer (v/v). For the cell lines listed in Table 1, four 10-cm cell culture dishes are optimal for lysosome enrichment. For other cell lines, cell density and the number of cell culture dishes may have to be adapted. During optimization, it is crucial to consider the binding capacity of the LS columns used for enrichment. For the cell lines and numbers indicated in Table 1, material obtained from up to eight 10-cm cell culture dishes can be enriched with one LS column.
9. The volume in which dounce homogenization is crucial for success. For a 15-mL douncer, a volume of 8–11 mL is recommended.
10. Fractions collected for β-hexosaminidase assays should not be frozen before the assay is performed. Freezing of these fractions leads to lysosomal membrane rupture and, hence, release of β-hexosaminidase, making it impossible to determine intact ratios.
11. The total fraction contains broken and intact lysosomes, cell debris, other organelles, cytosolic proteins, and intact cells.

12. Dounce-homogenization of the pellet after initial centrifugation can enhance lysosomal yield, as the pellet typically contains intact cells missed during the first homogenization. Another way to increase lysosomal yield is to pass the homogenate (without prior centrifugation) through a 23-gauge needle. Besides the dounce homogenizers, ball-bearing homogenizers can be utilized [9].
13. Re-application of the flow through to the column increases overall lysosomal yield as lysosomes that did not bind in the first passage can be retained.
14. LS columns contain iron particles, which may retain the magnetic field after removal from the magnetic stand. This can prevent efficient elution of lysosomes. Tapping of the column against a hard surface removes the residual magnetism.
15. Centrifugation of enriched lysosomal fractions at 20,000× *g* only pellets intact lysosomes, leaving fragments in solution. The resulting pellet is often difficult to dissolve without destroying lysosomes, possibly interfering with follow-up experiments.
16. In order to save material of the eluate fraction (enriched lysosomes), the input for the assay can be scaled down to a minimum of 5 μL. For low-abundance samples, it is recommended to increase the incubation time.

Acknowledgments

We would like to thank the Jürgen Manchot Stiftung for financial support in the course of a Ph.D. scholarship for Pathma Muthukottiappan.

References

1. Valm AM, Cohen S, Legant WR et al (2017) Applying systems-level spectral imaging and analysis to reveal the organelle interactome. Nature 546:162–167
2. Itzhak DN, Tyanova S, Cox J et al (2016) Global, quantitative and dynamic mapping of protein subcellular localization. elife 5:16950
3. Akter F, Bonini S, Ponnaiyan S et al (2023) Multi-cell line analysis of lysosomal proteomes reveals unique features and novel lysosomal proteins. Mol Cell Proteomics 22:100509
4. Singh J, Kaade E, Muntel J et al (2020) Systematic comparison of strategies for the enrichment of lysosomes by data independent acquisition. J Proteome Res 19:371–381
5. Walker MW, Lloyd-Evans E (2015) A rapid method for the preparation of ultrapure, functional lysosomes using functionalized superparamagnetic iron oxide nanoparticles. Methods Cell Biol 126:21–43
6. Singh J, Elhabashy H, Muthukottiappan P et al (2022) Cross-linking of the endolysosomal system reveals potential flotillin structures and cargo. Nat Comm 13:6212
7. Muthukottiappan P, Winter D (2021) A proteomic view on lysosomes. Mol Omics 17:842–859
8. Kyeong S, Kim J, Chang H et al (2021) Magnetic nanoparticles. Adv Exp Med Biol 1309: 191–215

9. Tharkeshwar AK, Trekker J, Vermeire W et al (2017) A novel approach to analyze lysosomal dysfunctions through subcellular proteomics and lipidomics: the case of NPC1 deficiency. Sci Rep 7:41408
10. Hamley IW (2003) Nanotechnology with soft materials. Angewandte Chemie-International Edition 42:1692–1712
11. Prousek J (2007) Fenton chemistry in biology and medicine. Pure Appl Chem 79:2325–2338
12. Duvvuri M, Krise JP (2005) A novel assay reveals that weakly basic model compounds concentrate in lysosomes to an extent greater than pH-partitioning theory would predict. Mol Pharm 2:440–448
13. Graham J (2002) Homogenization of mammalian cultured cells. Sci World J 2:794041
14. Winchester B (2005) Lysosomal metabolism of glycoproteins. Glycobiology 15:1r–15r
15. Wendeler M, Sandhoff K (2009) Hexosaminidase assays. Glycoconj J 26:945–952
16. Bonini S, Winter D (2024) Two-step enrichment facilitates background reduction for proteomic analysis of lysosomes. J Proteome Res 23:3393–3403

Chapter 8

Lysosomal Immunoprecipitation (LysoIP) from Cells and Tissues for Metabolomic and Proteomic Analyses

Nouf N. Laqtom and Monther Abu-Remaileh

Abstract

Lysosomes, known for degrading biomolecules and damaged cellular components, are now recognized as signaling hubs for nutrient sensing and metabolic adaptation, and their dysfunction is implicated in diseases including cancer and neurodegeneration. To understand the composition of the lysosome, the dynamic behavior of its contents, and its specific roles in health and disease, we describe a lysosomal immunoprecipitation method, termed "LysoIP," that enables the isolation of intact lysosomes from cultured cells and mouse tissues. This method utilizes a lysosome-localized 3xHA epitope tag (LysoTag) and a simple, yet robust organelle immunoprecipitation workflow. Isolated lysosomes are extracted with optimized buffers to allow the efficient retrieval of lysosomal proteins, polar metabolites, and lipids, maintaining compatibility with downstream liquid chromatography and mass spectrometry (LC-MS) analyses.

Key words Lysosomes, TMEM192, LysoTag, LysoTag mouse, LysoIP, LC-MS analyses, Metabolomics, Proteomics

1 Introduction

De Duve pioneered lysosome research with differential centrifugation, enabling the initial isolation and characterization of lysosomal content [1–3]. Later, Meyer and others refined organelle isolation with density gradient centrifugation [4–6]. While they significantly advanced the field, these methods often result in a final product contaminated by other organelles with similar densities, altered lysosomal lipid and metabolite profiles, and rupture or leakage from prolonged high-speed centrifugation, and they demand substantial starting material (reviewed in [7]). Recently, the development of a mitochondrial purification technique, mitoIP, inspired a similar approach for lysosomal immunoprecipitation [8, 9]. The lysosome immunoprecipitation (LysoIP) method relies on two key components: first, the tagging of a lysosomal transmembrane protein, TMEM192, with a 3xHA epitope on its cytosolic terminus as a

Stephanie M. Hughes and Indranil Basak (eds.), *Lysosomes: Methods and Protocols*, Methods in Molecular Biology, vol. 2976, https://doi.org/10.1007/978-1-0716-4844-5_8,

reliable marker; and second, magnetic beads conjugated with high-affinity HA-specific antibodies to selectively capture HA-tagged lysosomes [10]. For in vitro studies, stable LysoTag-expressing cell lines are generated, while for in vivo studies, a transgenic mouse line with Cre-loxP conditional activation enables TMEM192–3xHA expression in targeted tissues or cell types [10–12]. Upon tissue or cell disruption, a brief incubation with anti-HA magnetic beads efficiently captures TMEM192–3xHA-tagged lysosomes, which are then lysed to assess their contents. Purified lysosomes retain enzymatic activity, such as that of proteases and phospholipases [10, 12–14], and demonstrate metabolic intactness through the retention of lysosome-specific molecules (e.g., LysoTracker Red) and lysosome-enriched metabolites, like cystine, with minimal cytosolic contamination (e.g., absence of lactate) [10, 12]. LysoIP also facilitates tracing the degradation of labeled macromolecules taken up via endocytosis by monitoring labeled intermediates and end products [12, 15, 16]. Crucially, LysoIP uses buffers compatible with downstream multi-omics analyses by mass spectrometry (MS). Using this approach, we and others have successfully profiled the lysosomal proteome and metabolome in various cell lines and tissues under healthy, stress, and pathological conditions [10–13, 15–20].

2 Materials

2.1 Buffers for Metabolomics and Proteome Analysis

2.1.1 Washing Buffers for Metabolomics

Preparation of all buffers should be performed using LC-MS grade water and sterile, single-use 500-mL or 1000-mL PETG square bottles. Store the buffers at 4 °C to keep them cold and ready for experiments.

- 1× PBS (Phosphate-Buffered Saline): 136 mM NaCl, 2.7 mM KCl, 10 mM Na_2HPO_4, 1.8 mM KH_2PO_4, pH 7.4.
- 1× KPBS (Potassium Phosphate-Buffered Saline): 136 mM KCl, 10 mM KH_2PO_4, pH 7.25 (adjust with KOH as needed) (*see* **Note 1**).

2.1.2 Extraction Buffers for Metabolomics

For extraction buffers, use LC-MS grade methanol (MeOH), acetonitrile (ACN), isopropanol (IPA), and HPLC-grade stable chloroform ($CHCl_3$). Ensure all buffers are prepared and/or aliquoted in a clean hood designated for LC work to prevent contamination and maintain the purity of the reagents.

- Aliquot 250 mL of MeOH and 500 mL of $CHCl_3$ into 30 or 50 mL amber glass bottles. Store the aliquots in a −20 °C freezer. These aliquots will be used to prepare fresh chloroform-methanol (2:1) for lipid extraction.

- Use a new glass bottle to prepare 100 mL of a lipid resuspension buffer, consisting of 65% acetonitrile (65 mL), 30% isopropanol (30 mL), and 5% water (5 mL). Aliquot the prepared ACN-I-PA-H_2O buffer into 20-mL amber glass bottles. Store the aliquots at −20 °C.
- Use a new glass bottle to prepare 200 mL of 80% MeOH by mixing 160 mL of MeOH and 40 mL of water. Then, add 500 nM of isotopically labeled amino acids to serve as controls for extraction and mix thoroughly. Aliquot the prepared solution into 30- or 50-mL amber glass bottles. Store the aliquots at −20 °C. These aliquots will be used for polar metabolite extraction.

2.1.3 Washing Buffer for Proteome Analysis

- 1× PBS (Phosphate-Buffered Saline): 136 mM NaCl, 2.7 mM KCl, 10 mM Na_2HPO_4, 1.8 mM KH_2PO_4, pH 7.4. Commercially available PBS is sufficient, and LC-MS grade is not necessary. Store at 4 °C and before each use, aliquot the needed volume of this stock buffer and add protease inhibitors and phosphatase inhibitors if phosphorylated proteins are of interest.

2.1.4 Extraction Buffers for Proteome Analysis

- Lysis Buffer: 50 mM HEPES, pH 7.4 (adjusted with KOH), 40 mM NaCl, 2 mM EDTA, 1% Triton X-100, 1.5 mM sodium orthovanadate (Na_3VO_4), 300 mM sodium fluoride (NaF), 10 mM sodium pyrophosphate ($Na_4P_2O_7$), and 10 mM sodium β-glycerophosphate. Mix thoroughly and store the buffer at 4 °C for stability. Before each use, aliquot the needed volume of this stock buffer and add protease inhibitors and phosphatase inhibitors if phosphorylated proteins are of interest. This approach preserves protein integrity and phosphorylation status throughout the extraction process.
- Digestion buffer: 3 M urea in 0.1 M HEPES, pH 8.0. It should include 0.1 μg of lysyl endopeptidase C (LysC) added fresh before use to preserve enzyme activity. Store buffer without LysC in aliquots at −20 °C for up to 6 months.
- Solvent A consists of 0.05% formic acid in Milli-Q water, while Solvent B contains 80% acetonitrile and 0.05% formic acid in Milli-Q water; both should be prepared fresh or stored at room temperature for up to 1 week.
- The reconstitution buffer consists of 5% acetonitrile and 0.1% trifluoroacetic acid (TFA) in Milli-Q water, prepared fresh before use to ensure optimal sample quality.

2.2 LysoTag-Expressing Cell Line and LysoTag Mouse Model

- pLJC5-TMEM192-3xHA (Addgene #102930).
- pLJC6-TMEM192-3X-HA (Addgene #104434).
- psPAX2 (Addgene #12260) or pCMV-dR8.2 dvpr (Addgene #8455).

- pCMV-VSV-G (Addgene #8454).
- Polyethylenimine 25 K (PEI) (Polysciences, #23966-1).
- Polybrene.
- Puromycin.
- Blasticidin.
- LysoTag mouse model is available at JAX Laboratories (stock number 035401).
- Cre recombinase mouse line targeted for specific tissue or cell types of interest.
- Dissection tools.

2.3 Lysosomal Immunoprecipitation (LysoIP)

- LysoTracker Red DND-99.
- Anti-HA magnetic beads.
- Magnetic separation stand (DynaMag™- Spin Magnet).
- Vacuum for liquid aspiration (optional) (*see* **Note 2**).
- Sterile cell lifter with flat blade (for adherent cells).
- Refrigerated Centrifuge.
- Nutating Shaker (*see* **Note 3**).

  Prepare an appropriate cell disruption tool, as the choice may vary depending on the cell type. The following options are available and should be tested in this sequence:

  1. *Homogenizers*: Each consists of a 2-mL glass vessel (8 mm diameter, 120 mm height) with a serrated douncer that includes a TEFLON head and stainless steel shaft (*see* **Note 4**). This is the default tool for cell disruption unless the cell line requires a different approach (VWR, # 89026–386 and # 89026–398).
  2. *Cordless Motor-Driven Grinder*: If the homogenizer proves unsuitable for the cell line, use a cordless motor-driven grinder with clean, disposable pestles (DWK Life Sciences, #7495400000 and #K749521-1590).
  3. *Needle Method*: If the motor-driven grinder is also unsuitable, use a syringe fitted with a 27½ or 30½ gauge needle or a comfort point insulin syringe with a 29½ gauge needle.

3 Methods

3.1 Generation of Stable LysoTag Expressing Cell Line

3.1.1 For Viral Preparation

- Seed 1.5 million HEK293T cells in a 6-cm tissue culture dish.
- Prepare 1 μg of pLJC5-Tmem192-3xHA or pLJC6-3XHA-TMEM192, 900 ng of a lentivirus packaging plasmid (such as psPAX2 or dvpr), and 120 ng of VSV-G envelope plasmid. Mix the plasmids in 130 μL of DMEM (without serum) (*see* **Note 5**).

- Use Polyethylenimine 25 K (PEI) as the transfection agent at a 1:3 DNA to PEI ratio (for 2 μg of total DNA, use 6 μL of PEI (from a 1 mg/mL stock)). Vortex the PEI before adding it to the DNA mix.
- Incubate the DNA-PEI mixture at room temperature for 20 min.
- Transfect the cells by gently dropping the transfection mixture evenly across the dish. Shake the 6 cm dish in four directions to distribute the mixture.
- Incubate the cells at 37 °C in a 5% CO_2 tissue culture incubator.
- After 12–16 h, aspirate the media and replace it with fresh DMEM containing 30% IFS and 1% penicillin–streptomycin.
- Harvest the media containing the virus after 36 h and centrifuge to remove any floating cells.
- Aliquot the virus into 1 mL Aliquots in 1.7-mL Eppendorf tubes.
- Freeze the virus aliquots at −80 °C for at least 30 min to 1 h before use.

3.1.2 For Transduction

- Trypsinize the cells and count them.
- In a 6-well plate, sequentially add 10 μg/mL polybrene, media, cells, and the virus to make a total volume of 3 mL. For example, add 3 μL of polybrene (from a 10 mg/mL stock), 950 μL of media, 2 mL of cells (at 200,000 cells/mL), and 50 μL of virus (*see* **Note 6**). Also, reserve two wells without viruses to serve as controls for antibiotic selection.
- Centrifuge the plate at 2200× *g* for 45 min, then incubate at 37 °C with 5% CO_2.
- The following day, trypsinize the cells and transfer them to a 10-cm dish.
- After 24 h, begin antibiotic selection: For HEK cells infected with lentivirus carrying pLJC5-3xHA TMEM192-, use puromycin at a final concentration of 2 μg/mL. For cells infected with pLJC6-3XHA-TMEM192, use blasticidin at a final concentration of 10 μg/mL (*see* **Note 7**).
- Monitor the uninfected control wells for cell death, which should occur within 3 days for puromycin or 5 days for blasticidin.
- After completing selection, allow the cells to recover for 2 days, then continue growing them in the presence of a reduced concentration of antibiotics. For instance, use puromycin at a final concentration of 1 μg/mL and blasticidin at 5 μg/mL (*see* **Note 8**).
- Freeze some of these selected cells to avoid repeating the infection process for future experiments.

3.2 Lysosomal Immunoprecipitation (LysoIP) for Metabolomics from Cultured Cells

To maintain optimal conditions during the LysoIP experimental procedures, perform **steps 3** and **4** within a biosafety cabinet; **steps 5–11**, **16–21**, and **29–30** at the bench while keeping samples on ice; and **steps 12–15** and **22–27** in a cold room (*see* Fig. 1). For a seamless workflow, ensure all necessary reagents, buffers, consumables, and instruments are prepared in advance; refer to Checklist Table 1 for detailed guidance.

1. Grow LysoTag-expressing adherent cells in 15-cm cell culture dishes (avoid tissue culture flasks). Include untagged cells or cells labeled with a non-HA tag, such as TMEM192-2xFlag, as background controls. Ensure there are at least three biological replicates of each condition, along with an additional plate for cell counting and measuring total cell volume to enable accurate normalization and data processing. Once the cells reach 80% confluence, they are ready for the experiment.
2. Prepare a cooling rack (metal rack) on ice and arrange the required 2.0 mL V-bottom tubes for harvested cell suspension (CS), 1.5-mL tubes for homogenized suspensions (HS), whole-cell samples (WC), magnetic beads (MB), final LysoIP samples, and final WC samples. Label the final tubes with detailed sample information and experiment identifiers for accurate tracking.
3. Prepare sufficient volumes of plain media (media with no supplement) for washing and full media supplemented with 4 nM LysoTracker Red DND-99 for a one-hour incubation. To ensure sufficient volumes, include an additional 10 mL for each type of media. Each plate requires 10 mL of plain media for the initial wash and 10 mL of 4 nM LysoTracker Red DND-99 media for incubation (*see* **Note 9**). For example, if you have 8 plates, prepare 90 mL of plain media and 90 mL of LysoTracker-supplemented media.
4. Wash the first 15-cm dish with 10 mL of plain media, rotating the plate to ensure complete coverage of the cells. Leave for at least 30 s before aspirating the media. Meanwhile, proceed to wash the second plate. Replenish with media supplemented with 4 nM LysoTracker Red DND-99, and incubate the cells for one hour before proceeding with the next steps.

For optimal results, LysoIP should be conducted on two plates at a time. Plan the experiment with a clear, structured timeline, ensuring at least 20 min between replenishing each pair of plates. This staggering approach promotes smooth workflow and minimizes variations between samples.

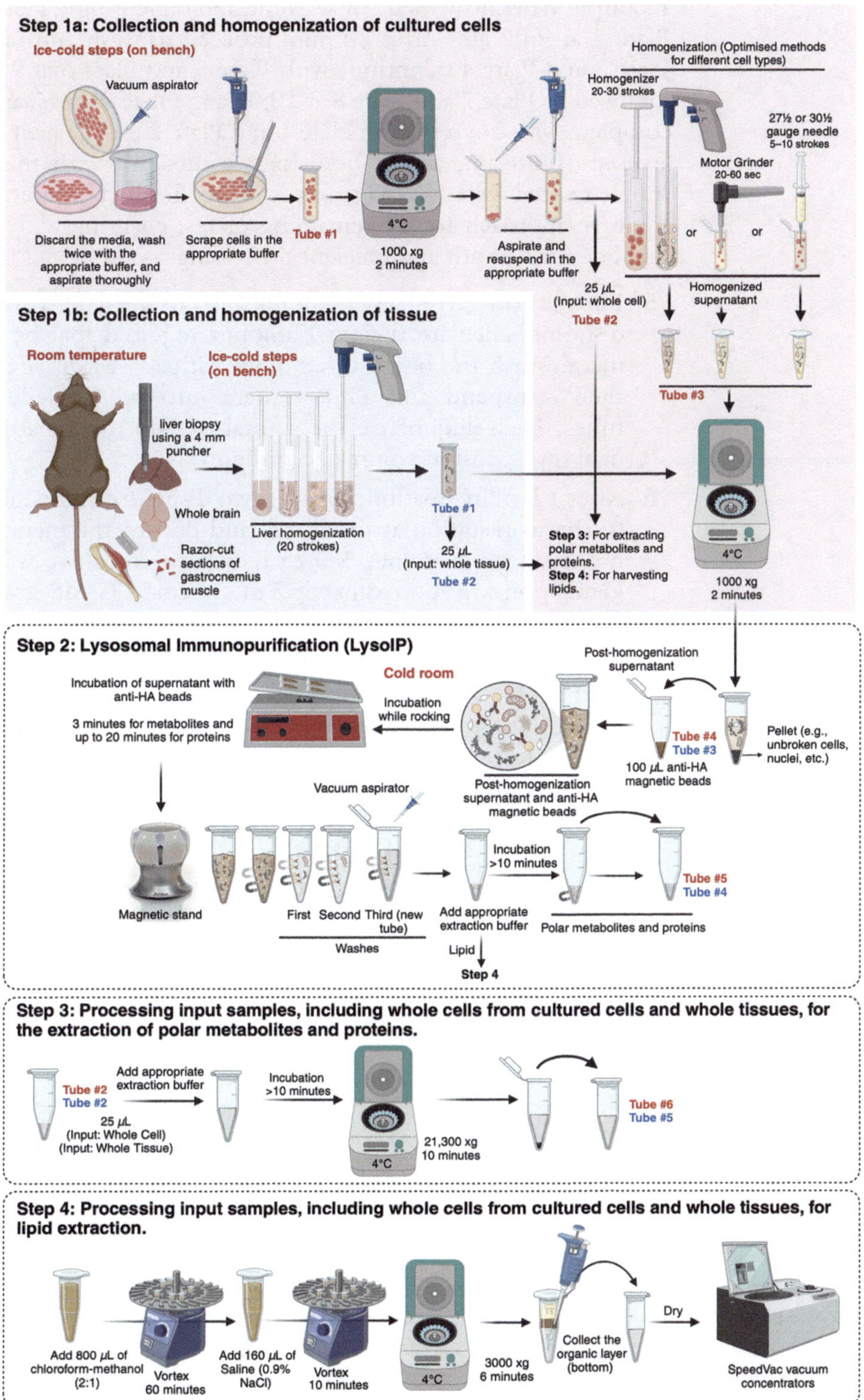

Fig. 1 Schematic diagram of the LysoIP workflow for isolating lysosomal polar metabolites, lipids, and proteins. Step 1 includes two distinct pathways: 1a (highlighted in red) for cultured cells and 1b (highlighted in blue) for mouse tissues. Steps 2, 3, and 4 are shared between both workflows. (This figure was generated using BioRender) BioRender.com (agreement number: KJ28NF9009)

Example Workflow Start by washing and replenishing Plate 1 and Plate 2 at 9:00 am. After 20 min, proceed to wash and replenish Plate 3 and Plate 4. Continue with Plate 5 and Plate 6 at 9:40 am, followed by Plate 7 and Plate 8 at 10:00 am. Once Plates 7 and 8 are complete, you can return to Plate 1 and Plate 2 for the next steps at around 10:00 am, as the hour-long incubation with the media supplemented with LysoTracker will be finished. Maintain this stepwise approach for all remaining plates, ensuring a consistent timeline for smooth and efficient processing.

5. For the anti-HA magnetic beads, plan to use 100–150 μL per dish and calculate the total amount required for the experiment. Wash the beads three times with LC-MS grade KPBS, then resuspend and aliquot them into pre-labeled, chilled tubes. Each aliquot can be a total of 100 μL, so adjust the final resuspension volume accordingly.
6. After 1 h of incubation, transfer two dishes from the incubator to the workstation at the bench and discard the media into a nearby waste container. For each dish, perform two washes by gently pouring approximately 5 mL of cold LC-MS-grade PBS to the edge of each 15 cm dish, discarding the wash PBS after each rinse, and finally aspirate to remove any residual PBS using a vacuum.
7. Following aspiration, add 950 μL of cold LC-MS-grade KPBS to the dish. Use a cell lifter to scrape the cells in a circular motion, then move from top to bottom to maximize cell collection efficiently. Transfer the collected cells into a 2-mL tube using a 1000-μL pipette.
8. Centrifuge the tubes at 4 °C and 1000× *g* for 2 min. Then aspirate to remove the KPBS using a vacuum, and resuspend the cell pellets in 950 μL of KPBS.

Fig. 1 (continued) Tube #1 (Red)—2-mL pre-homogenization tube: contains pre-homogenized cultured cells.
Tube #2 (Red)—1.5-mL input (whole cell) tube: contains whole cultured cells.
Tube #3 (Red)—1.5-mL cultured cell homogenization tube: contains homogenized tissue.
Tube #4 (Red)—1.5-mL anti-HA magnetic beads: contains anti-HA magnetic beads and homogenized cultured cells.
Tube #5 (Red)—1.5-mL final LysoIP tube: Final tube containing lysosomes from cultured cells.
Tube #6 (Red)—final cultured cell tube: final tube containing whole cultured cells.
Tube #1 (Blue)—2-mL tissue homogenization tube: contains homogenized mouse tissue.
Tube #2 (Blue)—1.5-mL input (whole tissue) tube: contains whole mouse tissue.
Tube #3 (Blue)—1.5-mL anti-HA magnetic beads: contain anti-HA magnetic beads and homogenized tissue.
Tube #4 (Blue)—1.5-mL final LysoIP tube: final tube containing lysosomes from tissue.
Tube #5 (Blue)—1.5-mL final whole tissue tube: final tube containing whole tissue.

Table 1
Checklist for smooth workflow

Workstations	Items	Details
Biosafety Cabinet	Plain Media	10 mL per tissue culture dish (no supplements)
	Media with LysoTracker Red DND-99	10 mL per dish if harvesting polar metabolites; omit LysoTracker for protein or lipid extraction
Bench	*Ice Bucket*	
	LC-MS grade PBS	10 mL per dish (polar metabolites and lipids)
	LC-MS grade KPBS	2 mL per dish (polar metabolites and lipids)
	PBS	10 mL per dish (protein analysis)
	PBS with protease/phosphatase inhibitors	2 mL per dish (protein analysis)
	1.5 mL and 2 mL tubes	Five 1.5-mL tubes and one 2-mL tube per dish; pre-label and keep on ice
	Instruments	
	Vacuum for liquid aspiration	Pre-set to 2 min at 1000× *g* (only for cold centrifuge)
	Cold centrifuge	
	Magnetic separation stand	Pre-set to 3 min (only for the timer)
	Timer	
	Appropriate cell disruption tool:	
	Homogenizers	2-mL glass vessel with serrated douncer. Recommended for HEK293 and HeLa
	Cordless motor-driven grinder	Clean, disposable pestles; fresh pestle per dish Recommended for RPE-19 and PaTu 8988 t
	Needle method	27½ or 30½ gauge needle per dish. Recommended for LCLs and SH-SY5Y
	Consumables	
	Cell scraper	One per dish.
	1 mL serological pipette	One per dish.
	Pipettes	Pre-set to 25 μL and 950 μL.
	Pipette tips	200-μL and 1000-μL tips
	Waste disposal	Liquid waste container and solid waste bin

(continued)

Table 1
(continued)

Workstations	Items	Details
Cold Room	*Buffers*	
	LC-MS-grade KPBS	3 mL per dish (polar metabolites and lipids)
	Extraction buffer for polar metabolites	80% methanol with isotopically labeled amino acids (300 μL per dish)
	Lipid extraction buffer	Chloroform: Methanol (2:1) with isotope-labeled lipidomics standard (0.75 μg/mL) (2 mL per dish)
	PBS with protease/phosphatase inhibitors	3 mL per dish (protein analysis)
	Protein lysis buffer	Freshly add protease/phosphatase inhibitors (240 μL per dish)
	Instruments	
	Vacuum for liquid aspiration (optional)	
	Magnetic separation stand	
	Consumables	
	Pipettes	50 μL and 950 μL (polar metabolites); 950 μL and 800 μL (lipids)
	Tubes	1.5-mL tubes for the final wash step
	Low retention tips	200-μL and 1000-μL tips
	Waste bin	For solid waste disposal

The checklist ensures all materials and equipment are organized and ready for use before starting the experiment. This helps optimize the workflow and reduce potential disruptions

9. Transfer 25 μL of the cell suspension into a pre-labeled tube as the input (whole cell) sample, representing 2.5% of the total cells.
10. Transfer the remaining suspension into a clean, pre-chilled 2-mL glass vessel and disrupt the cells by homogenizing with a douncer, performing 20–30 strokes. For some cell lines, a cordless motor-driven grinder may offer more efficient disruption within 30–60 seconds, or use a syringe fitted with a 27½ or 30½ gauge needle, applying 5–10 strokes (*see* **Note 10**).
11. Use a 1-mL serological pipette to transfer the cell homogenate from the glass vessel into a 1.5-mL tube (*see* **Note 11**). Then, centrifuge the tube at 4 °C and 1000× *g* for 2 min (*see* **Note 12**).
12. Gently collect the organelle-containing homogenate without disturbing the pellet, and immediately transfer it to 100 μL of pre-chilled, washed anti-HA magnetic beads.
13. Place the tube on a rocker in the cold room and shake for 3 min to allow the anti-HA beads to capture the HA-tagged lysosomes.
14. Place the tube on a magnetic stand to separate the lysosomes, which are captured by the magnetic beads, from the remaining homogenate, including other organelles and cytosolic contents. Carefully aspirate and discard the homogenate.
15. Remove the tube from the magnetic stand and wash the beads with 950 μL of KPBS to eliminate any nonspecific binding. Repeat this wash step at least three times. For the final wash, transfer the beads to a fresh, clean tube (*see* **Note 13**). The subsequent steps will depend on whether processing for polar metabolites (**steps 16–22**) or lipid metabolites (**steps 23–31**).
16. *For polar metabolite extraction*, resuspend the beads in 50 μL of 80% methanol spiked with 500 nM isotopically labeled amino acids (*see* **Note 14**). Keep all samples in the cold room until all plates are processed, allowing the final sample to sit for an additional 10–15 min.
17. During this 10–15 min, bring the input (whole cell) samples to the cold room, add 225 μL of methanol solution for metabolites, and vortex for 10 s. Centrifuge at 21,300× *g* (maximum speed) for 10 minutes, then transfer the input sample to clean, pre-chilled tubes.
18. Use a magnetic stand to separate the methanol extraction buffer containing the lysosomal metabolites from the beads, then transfer the supernatant to pre-chilled tubes.
19. For solvent and buffer controls, prepare one tube with 25 μL of KPBS buffer (as used in this experiment) topped with 225 μL

of methanol extraction buffer. Prepare a second tube with 50 μL of extraction buffer alone.

20. Store both the input (whole cell) and lysosomal samples at −80 °C.

21. To prepare for mass spectrometry analysis, vortex the samples for 10 min in the cold room, then centrifuge at 21,300× *g* (maximum speed) for an additional 10 minutes. Transfer the required volume of supernatant into the V-shaped LC-MS vials for analysis. Shake to ensure the sample settles at the bottom, confirming there are no air bubbles.

22. In addition to individual samples, prepare a pooled QC sample by combining a few μL from each input sample to create a pooled whole-cell sample, and a few μL from each lysosomal fraction to create a pooled LysoIP sample. Use this pooled QC sample to prepare dilutions for ensuring linearity in mass spectrometry analysis. Start with the undiluted pooled sample as 1×, then prepare dilutions at 0.6×, 0.3×, and 0.1×, or alternatively at 0.5× and 0.1×. (*see* **Note 15**).

23. *For lipid extraction*, resuspend the beads from **step 15** in 800 μL of chloroform–methanol (2:1) with 0.75 μg/mL isotope-labeled lipid standards (*see* **Note 14**). Keep all samples in the cold room until all plates are processed, allowing the final sample to sit for an additional 10–15 min (*see* Fig. 1).

24. During this 10–15 min period, bring the input (whole cell) samples to the old room and add 800 μL of chloroform–methanol (2:1) solution for lipids.

25. Prepare solvent and buffer controls by setting up one tube with 25 μL of KPBS buffer (as used in this experiment) and 800 μL of chloroform–methanol (2:1) extraction buffer. Also, prepare a second tube containing 800 μL of extraction buffer alone.

26. Vortex all samples for 1 h.

27. Add 160 μL (1/5 volume) of 0.9% saline solution to each sample and vortex for 10 min in the cold room.

28. Centrifuge at 3000× *g* for 5 min. Then, transfer the organic phase (bottom layer) containing the lysosomal nonpolar content into clean, pre-chilled tubes. Ensure consistent volume transfer across samples (e.g., 550 or 600 μL).

29. Dry lipid samples in a speed vacuum at room temperature using a high vacuum setting. Dry the samples for 45–120 min.

30. Store both the input (whole cell) and lysosomal samples at −80 °C. To prepare for mass spectrometry analysis, freshly prepare an ACN-IPA-H_2O buffer (65:30:5) and add 50 μL of this buffer to each sample. Vortex for 10 min in the cold room, then centrifuge at 21,300× *g* (maximum speed) for an

additional 10 min. Carefully transfer the required volume of supernatant into glass LC-MS vials for analysis.

31. In addition to individual samples, prepare a pooled QC sample by taking a few μL from each input sample to create a pooled whole-cell sample and a few μL from each lysosomal fraction to create a pooled LysoIP sample (*see* **Note 15**). As described in **step 22**, prepare dilutions from these pooled samples as well.

3.3 Lysosomal Immunoprecipitation (LysoIP) for Protein Analysis from Cultured Cells

Follow the same steps as outlined in the method above, unless otherwise specified (*see* Fig. 1).

1. Wash the cells with 10 mL of plain media and then replenish with 10 mL of fresh media, without adding LysoTracker, as it will not be detectable during proteomic analysis.
2. After 1 h of incubation, wash each dish twice using 5 mL of cold PBS. Add 950 μL of cold PBS supplemented with protease and phosphatase inhibitors (if phosphorylation analysis is of interest) to each dish. Use a cell lifter to gently scrape the cells, then transfer the collected cells into a 2-mL tube.
3. Centrifuge the tubes, aspirate the PBS, and resuspend the cell pellets in 950 μL of PBS supplemented with inhibitors.
4. Transfer 25 μL of the suspension as the input (whole cell) sample, and proceed to homogenize the remaining suspension using the appropriate cell disruption method (*see* **Note 10**).
5. Centrifuge the homogenate of disrupted cells at 4 °C and 1000× *g* for 2 min.
6. Collect the homogenate without disturbing the pellet and incubate it with 100–150 μL of pre-chilled, washed anti-HA magnetic beads for 3–15 min with gentle shaking in a cold room. Wash the anti-HA magnetic beads three times with PBS supplemented with protease inhibitors (*see* **Note 16**).
7. Extract proteins using 80 μL of lysis buffer (*see* Subheading 2.1.4) for LysoIP samples and 160 μL for input (whole-cell) samples.
8. *For Western blot analysis*, immediately add 1× SDS loading dye to the samples. The samples are now ready to load onto the Western blot gel. For the best results, load 2× the volume of lysosomal samples; for example, if using 10 μL of input (whole cell), load 20 μL of lysosomal samples. Avoid normalizing protein concentration, as lysosomal protein samples are typically below or near the background levels.
9. *For proteomics analysis* (*see* **Note 17**), add 20% (w/v) SDS to the sample to achieve a final concentration of 2%, then sonicate using the Bioruptor Plus (Diagenode) for 5 cycles (1 min on, 30 s off, at 20 °C, highest setting).

10. Centrifuge the samples at 21,300× *g* (maximum speed) for 1 min and transfer the supernatant to fresh tubes. Add 2.9 μL of 200 mM dithiothreitol (DTT) per 20 μL of protein sample for reduction. Mix gently and incubate for 15 min at 45 °C.
11. For alkylation, add 5 μL of 200 mM iodoacetamide (IAA) per 20 μL of reduced protein sample. Mix thoroughly and incubate for 30 min at room temperature in the dark. Optionally, add an additional 2.9 μL of 200 mM DTT to quench excess IAA before proceeding.
12. Precipitate proteins by adding 4 volumes of ice-cold acetone (e.g., for a 20-μL protein sample, add 80 μL of acetone). Mix gently and incubate overnight at −20 °C.
13. Centrifuge the precipitated proteins at 21,300× *g* (maximum speed) for 30 min at 4 °C. Carefully remove the supernatant without disturbing the pellet. Wash the pellet twice with 500 μL of 80% (v/v) ice-cold acetone. After each wash, vortex briefly and centrifuge at 21,300× *g* (maximum speed) for 2 min at 4 °C.
14. Air-dry the pellet until no residual acetone remains (typically 5–10 min). Avoid overdrying, as this may hinder resuspension in subsequent steps.
15. Resuspend the air-dried pellet in 50 μL of digestion buffer containing 0.1 μg of lysyl endopeptidase (LysC) per 50 μg of protein (*see* Subheading 2.1.4). Incubate for 4 h at 37 °C with shaking at 600 rpm.
16. Dilute the sample 1:1 with Milli-Q water (e.g., add 50 μL of Milli-Q water to 50 μL of digested sample) to reduce the urea concentration to 1.5 M. Add 0.1 μg of trypsin per 50 μg of protein and incubate for 16 h at 37 °C.
17. Acidify the digested peptides with 10% trifluoroacetic acid (TFA) by adding 5 μL of TFA per 50 μL of digested sample. Condition the Waters Oasis HLB μElution Plate by adding 100 μL of solvent B (*see* Subheading 2.1.4) three times, allowing each addition to flow through before the next.
18. Equilibrate the plate by adding 100 μL of solvent A (*see* Subheading 2.1.4) three times, allowing each addition to flow through. Load the acidified peptide samples onto the plate under slow vacuum.
19. Wash the plate by adding 100 μL of solvent A three times to remove salts and impurities, then elute the peptides with 50 μL of solvent B into PCR tubes. After elution, process the samples immediately or store them at −80 °C to prevent degradation.
20. Dry the eluted peptides in a SpeedVac and resuspend them in 10 μL of reconstitution buffer (*see* Subheading 2.1.4). Vortex or shake the tubes at 1500–2000 rpm at room temperature to ensure proper reconstitution.

21. Centrifuge the peptides at 21,300× *g* (maximum speed) for 10 min and transfer the supernatant to an HPLC vial for LC-MS/MS analysis.

3.4 LysoIP for Metabolomic and Proteomic Analyses from Mouse Tissues

Our LysoTag mouse model utilizes a Cre-lox system, requiring crosses between LysoTag mice and Cre-expressing mice of interest to achieve tissue-specific lysosomal tagging. For experiments, ensure 4–5 LysoTag$^{+/-}$ mice are included for each experimental genotype or condition, with at least three LysoTag$^{-/-}$ mice as control background (*see* **Note 18**).

Process one animal at a time if harvesting 2–3 tissues, or up to two animals if harvesting a single tissue. Schedule the experiment with at least a 30-min interval between each set.

1. Euthanize by cervical dislocation or the method approved in your protocol, then immediately dissect to harvest the tissues of interest. Optimize the tissue amounts required for analysis. Based on our work, a 3-mm biopsy punch can be used for the liver, while the entire cerebral cortex should be collected for brain samples, and the whole heart should be taken and cut into smaller pieces with a razor blade before homogenization.
2. Prepare a clean, pre-chilled 2-mL glass vessel on ice. For metabolomics, fill the vessel with 1 mL of LC-MS-grade KPBS buffer; for protein analysis, use 1 mL of PBS supplemented with protease and/or phosphatase inhibitors. Place the tissue in the vessel and homogenize with 20–30 strokes. Collect the homogenate using a 1-mL serological pipette and transfer it to a 1.7-mL tube.
3. Transfer 25 μL of the homogenate as the input (whole cell) sample, and proceed to centrifuge the remaining at 4 °C and 1000× *g* for 2 min.
4. Collect the homogenate without disturbing the pellet and incubate it with 100–150 μL of pre-chilled, washed anti-HA magnetic beads.
5. Follow **steps 13–31** in the LysoIP protocol for metabolomics from cultured cells. For protein analysis, follow from **step 6** onward.

4 Notes

1. Measure pH using pH strips, avoiding direct insertion of objects into the KPBS solution. Instead, transfer a few drops of the buffer into a clean tube and measure the pH. Prepare the pH adjuster, KOH, with LC-MS grade water to maintain the purity and consistency of the buffer.

2. Using a vacuum for liquid aspiration, though optional, can speed up washing steps, reduce residual liquid, and minimize pipette tip waste compared to manual methods.
3. Set the nutating shaker to a fixed speed of 60–70 rpm during the experiment, or use a model with a preset fixed speed to ensure consistent mixing.
4. Consider using a larger glass vessel, such as 5 mL, 7 mL, or 15 mL, for large cell quantities (e.g., over 50 million cells) or larger tissue samples to accommodate the sample volume and ensure effective disruption.
5. Use endotoxin-free purification kit for plasmid preparation to maintain high-quality viral production and reduce cytotoxicity in transduced cells. The choice between using pLJC5-TMEM192-3xHA or pLJC6-3XHA-TMEM192 typically depends on the antibiotic resistance already present in the cells. If the cells are already resistant to puromycin, it is preferable to use pLJC6-3XHA-TMEM192, which carries a blasticidin resistance marker, to allow for effective selection.
6. Testing viral stock helps determine the appropriate amount needed to achieve approximately 50% infection efficiency. This test helps avoid overexpression of the lysosomal tag in transduced cells, which could result in contamination from other organelles due to excessive tagging.
7. Perform antibiotic concentration testing, as different cell lines may require adjustments for optimal selection. Setting up a plate with varying antibiotic concentrations allows the identification of the concentration that efficiently kills uninfected cells within 3–5 days. This method ensures that the optimal selection is applied, allowing the removal of non-transduced cells while avoiding cytotoxic effects on transduced cells.
8. Maintaining cells in a reduced concentration of antibiotics ensures a low level of selection pressure, allowing for the elimination of cells that no longer express the LysoTag. Meanwhile, cells that continue to stably express the LysoTag will proliferate.
9. LysoTracker is crucial for estimating the efficiency of LysoIP and ensuring that an equivalent amount of lysosomes is immunoprecipitated across samples. As a small fluorescent molecule, it is detectable by mass spectrometry; therefore, microscopic examination is not needed to confirm lysosomal labeling.
10. The method of cell disruption is critical to optimize when working with a cell line for the first time. It's essential to be gentle enough to break the cell membranes without damaging the organelles, which could lead to leakage or loss of their contents.

11. It is important to use a 1-mL serological pipette, as it is the best fit to reach the bottom of the vessel. Using other serological pipettes or a 1000-μL tip may not provide a proper fit and could compromise the harvesting of the homogenate.
12. Washing the vessel and douncer of the homogenizer between culture plates requires filling the vessel with Milli-Q water, flicking it to rinse, and repeating this process 10 times before placing the vessel upside down to air-dry. For the douncer, hold it by the handle and rinse it under running Milli-Q water, moving it up and down 10 times to ensure thorough cleaning, maintaining consistency in this washing method for the homogenizer. For tissue samples, begin with an initial rinse using tap water to remove any tissue debris stuck to the vessel walls, followed by 10 rinses with DDI water as described. This cleaning step can be performed during the two-minute centrifugation mentioned in **step 11**.
13. Transfer the beads to a new tube during the final wash to minimize potential contaminants adhering to the walls of the original tube.
14. Stable isotope-labeled standard used for internal calibration in mass spectrometry (MS) analyses.
15. Exclude all background samples when pooling the LysoIP samples.
16. Incubating the homogenate with anti-HA magnetic beads for up to 15 min enhances lysosomal protein yield; however, it does not improve metabolite yield.
17. It is recommended to use low-protein-binding tubes (LoBind, Eppendorf) for all steps in proteomics experiments to minimize variability between samples and ensure maximum protein recovery.
18. Use a single allele for the LysoTag to prevent overexpression and minimize potential artifacts from excessive tagged protein levels.

Acknowledgments

We sincerely thank all members of the Laqtom and Abu-Remaileh Labs for their invaluable discussions and insightful contributions to this work.

References

1. De Duve C, Beaufay H (1981) A short history of tissue fractionation. J Cell Biol 91(3 Pt 2): 293s–299s
2. De Duve C et al (1955) Tissue fractionation studies. 6. Intracellular distribution patterns of enzymes in rat-liver tissue. Biochem J 60(4): 604–617
3. Leighton F et al (1968) The large-scale separation of peroxisomes, mitochondria, and lysosomes from the livers of rats injected with triton WR-1339. Improved isolation procedures, automated analysis, biochemical and morphological properties of fractions. J Cell Biol 37(2):482–513
4. Henning R, Plattner H (1974) Isolation of rat liver lysosomes by loading with colloidal gold. Biochim Biophys Acta 354(1):114–120
5. Rickwood D, Birnie GD (1975) Metrizamide, a new density-gradient medium. FEBS Lett 50(2):102–110
6. Symons LJ, Jonas AJ (1987) Isolation of highly purified rat liver lysosomal membranes using two Percoll gradients. Anal Biochem 164(2): 382–390
7. Chen C, Sidransky E, Chen Y (2022) Lyso-IP: uncovering pathogenic mechanisms of lysosomal dysfunction. Biomolecules 12(5)
8. Bayraktar EC et al (2019) MITO-Tag Mice enable rapid isolation and multimodal profiling of mitochondria from specific cell types in vivo. Proc Natl Acad Sci USA 116(1):303–312
9. Chen WW et al (2016) Absolute quantification of matrix metabolites reveals the dynamics of mitochondrial metabolism. Cell 166(5): 1324–1337. e11
10. Abu-Remaileh M et al (2017) Lysosomal metabolomics reveals V-ATPase- and mTOR-dependent regulation of amino acid efflux from lysosomes. Science 358(6364):807–813
11. Ghoochani A et al (2024) Cell-type resolved protein atlas of brain lysosomes identifies SLC45A1-associated disease as a lysosomal disorder. bioRxiv
12. Laqtom NN et al (2022) CLN3 is required for the clearance of glycerophosphodiesters from lysosomes. Nature 609(7929):1005–1011
13. Nyame K et al (2024) Glycerophosphodiesters inhibit lysosomal phospholipid catabolism in Batten disease. Mol Cell 84(7):1354–1364 e9
14. Pedram K et al (2022) Lysosomal cathepsin D mediates endogenous mucin glycodomain catabolism in mammals. Proc Natl Acad Sci USA 119(39):e2117105119
15. Medoh UN et al (2023) The Batten disease gene product CLN5 is the lysosomal bis (monoacylglycero)phosphate synthase. Science 381(6663):1182–1189
16. Scharenberg SG et al (2023) An SPNS1-dependent lysosomal lipid transport pathway that enables cell survival under choline limitation. Sci Adv 9(16):eadf8966
17. Berquez M et al (2023) Lysosomal cystine export regulates mTORC1 signaling to guide kidney epithelial cell fate specialization. Nat Commun 14(1):3994
18. Gupta S et al (2021) Lysosomal retargeting of Myoferlin mitigates membrane stress to enable pancreatic cancer growth. Nat Cell Biol 23(3): 232–242
19. Hundley FV et al (2024) Endo-IP and lyso-IP toolkit for endolysosomal profiling of human-induced neurons. Proc Natl Acad Sci USA 121(52):e2419079121
20. Yamamoto K et al (2020) Autophagy promotes immune evasion of pancreatic cancer by degrading MHC-I. Nature 581(7806): 100–105

Chapter 9

Mass Spectrometry-Based Proteomic Analysis of Lysosome-Enriched Fractions

Sara Bonini, Pathma Muthukottiappan, and Dominic Winter

Abstract

Lysosomes are the main degradative organelles of most mammalian cells, playing a crucial role in nutrient metabolism and acting as a signaling hub. Due to the low abundance of lysosomes, several enrichment techniques have been developed to facilitate their investigation. Combination of these approaches with mass spectrometry-based proteomics presents a powerful tool for the unbiased characterization of the lysosomal proteome under various conditions. Different enrichment strategies result in lysosome-enriched fractions with unique characteristics, varying, for example, in protein concentration and buffer composition. Therefore, in order to obtain optimal results, the subsequent strategy for proteomics sample preparation has to be adapted. Here, we describe different methods for the processing of lysosome-enriched fractions for mass spectrometry-based proteomic analyses and provide guidance for the selection of the appropriate strategy. We further elaborate on typical parameters for mass spectrometric analysis and data processing.

Key words Proteomics, Mass spectrometry, Lysosomes, Urea, RapiGest, SP3

1 Introduction

To date, >350 proteins have been assigned with high confidence to be located at lysosomes, or to be important for their function, and this list is continuously growing [1, 2]. The main tool utilized for the characterization of the lysosomal proteome under different cellular conditions and disease pathologies, as well as for the identification of novel lysosomal proteins, is mass spectrometry (MS)-based proteomics, as it enables the simultaneous identification and quantification of thousands of proteins from a given sample [1–5]. Such MS analyses are typically performed using the so-called bottom-up approach, which is based on the enzymatic digestion of the samples' proteins, followed by the liquid chromatography tandem MS (LC-MS/MS) analysis of the resulting peptides [6]. Key steps in this context are the purification and denaturation of proteins, the reduction/alkylation of their disulfide bonds [7], the

Stephanie M. Hughes and Indranil Basak (eds.), *Lysosomes: Methods and Protocols*, Methods in Molecular Biology, vol. 2976, https://doi.org/10.1007/978-1-0716-4844-5_9,

proteolytic digestion of proteins, the desalting of the resulting peptides, and finally, their analysis by LC-MS/MS [8]. For the latter, it is of crucial importance that the peptides originating from the proteins of interest are reproducibly detected with sufficient signal intensity by the mass spectrometer, enabling their quantitative analysis with high performance. In this context, the investigation of lysosomal proteins is often challenging, as their low abundance (~0.2% of cellular mass [9]) often results in an irreproducible/unreliable detection in untargeted analyses of whole cell/ tissue lysates [1, 10, 11].

Lysosome enrichment techniques, which result in a decrease of general sample complexity, and hence an increase in signal intensity for lysosomal protein-derived peptides, present an attractive solution to this problem and have been utilized in numerous proteomics studies (reviewed in [2]). These approaches allow not only for the sensitive and robust characterization of the lysosomal proteome but also for the identification of novel lysosomal proteins [1, 3]. There are several techniques for the enrichment of lysosomes, of which sucrose density gradient centrifugation [3, 12], superparamagnetic iron-oxide nanoparticles (SPIONs) [13], and immunoprecipitation of tagged lysosomal membrane proteins (Lyso-IP) [5] are currently used most frequently.

With respect to subsequent sample preparation strategies for LC-MS/MS analysis, the lysosome-enriched fractions obtained by the individual approaches differ in several parameters, such as the protein concentration, total protein amount, or buffer composition, affecting the decision which strategy is best-suited. For example, the presence of MS-incompatible chemicals (such as detergents) might result in the need for protein purification steps, while low amounts of protein complicate protein precipitation and might require specific protocols to avoid sample loss. Here, we present a modular workflow for the MS sample preparation of lysosome-enriched fractions, which can be tailored toward the characteristics of the respective enrichment approach.

2 Materials

All buffers should be prepared with MS-grade water, and fresh gloves and plasticware should be used to avoid contamination with detergents or skin/hair proteins. Buffers should not be sterile filtered, as this often results in polymer contamination. Moreover, low protein binding tubes and pipette tips are highly recommended to minimize sample loss.

2.1 Protein Purification and Digestion

1. Acetone.
2. Vortex mixer.
3. Bench-top refrigerated centrifuge.
4. Methanol.
5. Chloroform/methanol, mixed in a 2:1 (v/v) ratio.
6. Glass syringe.
7. Trichloroacetic acid (TCA) 100% (w/v) solution (*see* **Notel**).
8. Urea, 8 M stock solution in 0.1 M TEAB buffer (*see* **Note 2**).
9. Tetraethylammonium bicarbonate (TEAB), pH 8.5, 1 M stock solution (*see* **Note 3**).
10. Thermomixer.
11. Dithiothreitol (DTT), 1 M stock solution (*see* **Note 4**).
12. Acrylamide or chloroacetamide [14], 2 M stock solution (*see* **Note 4**).
13. rLysC, MS-grade, 0.1 μg/μL stock solution (*see* **Note 5**).
14. Trypsin, MS-grade, 0.1 μg/μL stock solution (*see* **Note 5**).
15. Acetic acid (AA).
16. Formic acid (FA).
17. Trifluoroacetic acid (TFA).
18. RapiGest, 10% stock solution in water (*see* **Note 6**).
19. Ammonium bicarbonate (ABC, NH_4HCO_3) pH 7.8, 0.1 M stock solution.
20. Ultrasonic water bath.
21. Sera-Mag SpeedBeads mixed in a 1 to 1 ratio [15] (*see* **Note** 7).
22. PCR tubes.
23. Magnetic stand for PCR tubes.
24. Ethanol 96% and 80%.

2.2 Desalting of Peptides

1. SPE Vacuum Manifold (*see* **Note 8**).
2. Oasis HLB 1 cc Vac Cartridge, 10 mg sorbent per cartridge, 30 μm particle size.
3. Elution Buffer: 80% acetonitrile (ACN), 0.5% AA.
4. Washing Buffer: 0.5% acetic acid (AA).
5. Sep-Pak reservoir adapter.
6. Disposable syringes (10 mL volume).
7. Vacuum centrifuge.
8. Resuspension Buffer 1: 5% ACN.
9. Resuspension Buffer 2: 5% ACN 5% FA.

10. Fluorometric Peptide Assay (commercially available kit from Thermo Scientific).
11. C_{18} Octadecyl reversed phase extraction disks (*see* **Note 9**).
12. Methanol.
13. Bench-top centrifuge.

2.3 LC-MS/MS Analysis

1. Analytical column: Fused silica capillary with 360 μm outer diameter (OD), 100 μm inner diameter (ID), packed with C_{18} reversed phase 3 μm particles or a similar column.
2. Dionex Ultimate 3000 nanoflow ultra-high-performance liquid chromatography (UHPLC) device or a similar nanoflow liquid chromatography system.
3. Buffer A: 0.1% FA.
4. Buffer B: 90% ACN, 0.1% FA.
5. Orbitrap Fusion Lumos or a similar mass spectrometer.
6. Spectronaut or similar software for mass spectrometry data analysis.

3 Methods

Depending on the amount of protein present in the lysosome-enriched fraction, different approaches can be applied (Fig. 1). This is mainly due to the protein concentration, and hence the possibility of precipitating proteins. While all samples can be processed with single-pot, solid-phase-enhanced sample-preparation (SP3), protein precipitation can only be performed with high efficiency in samples exceeding a certain protein concentration (*see* **Note 10**).

3.1 Protein Precipitation

In particular, samples derived from Lyso-IP experiments, which require elution of lysosomal proteins with detergents such as Triton X-100 or sodium dodecyl sulphate (SDS), require a subsequent step to separate proteins from detergents, which would interfere with MS analysis. Acetone precipitation is a fast and robust approach, which is suitable for samples with moderate protein and salt concentrations. In the case of samples with high protein amount and low salt concentration, chloroform/methanol precipitation offers superior results as it results in a protein disk, which is easy to dissolve, and lipids are efficiently sequestered in the chloroform phase. When the sample is low in protein concentration and/or contains a high amount of salts, on the other hand, TCA performs best. For highly diluted samples, or total protein amounts below 50 μg, it is not advised to perform a precipitation step, since the resulting protein pellet will be too small. In this case, it is advisable to process the sample by SP3.

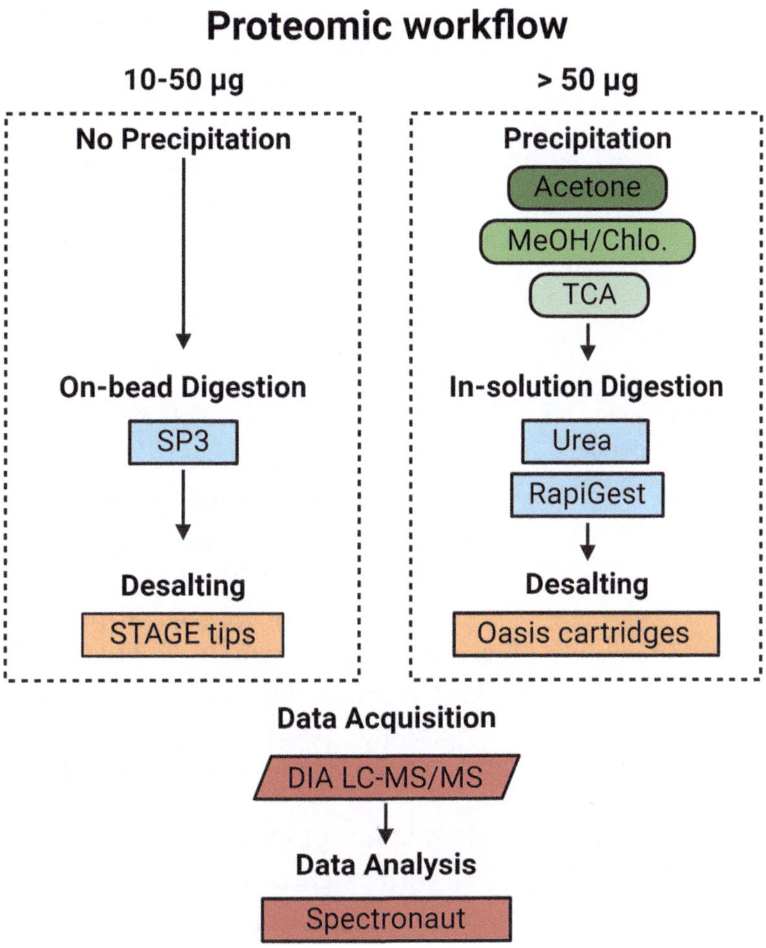

Fig. 1 Decision tree for the selection of workflows to prepare lysosome-enriched samples for proteomic analysis based on the amount of protein input. For samples containing higher protein amounts (> 50 μg), protein precipitation using acetone, chloroform/methanol, or TCA presents a suitable option to remove interfering substances. Subsequently, the precipitated proteins can be digested in solution using either urea or RapiGest. Alternatively, in the case of lower protein amounts, which are insufficient/impractical for precipitation (typically < 50 μg), digestion using SP3 is recommended. For peptide desalting, either STAGE tips or Oasis cartridges can be applied, based on the amount of sample. Peptide samples are analyzed using DIA-LC-MS/MS, and protein identification/quantification is performed using Spectronaut. MeOH methanol, Chlo. chloroform, TCA trichloroacetic acid, SP3 single-pot, solid-phase-enhanced sample-preparation, STAGE stop and go extraction, DIA data independent acquisition, LC-MS/MS liquid chromatography-tandem mass spectrometry

3.1.1 Acetone Precipitation

1. Add 4× the sample volume of ice-cold acetone to the sample.
2. Vortex vigorously for 15 s.
3. Incubate at −20 °C for at least 3 h (*see* **Note 11**).

4. Centrifuge the sample at 20,000× *g* for 30 min at 4 °C (*see* **Note 12**).
5. Gently aspirate and discard the supernatant without disturbing the pellet.
6. Add 1 mL of ice-cold methanol to the sample and briefly vortex to detach the pellet from the bottom of the tube.
7. Centrifuge the sample at 20,000× *g* for 15 min at 4 °C.
8. Gently aspirate the supernatant and air-dry the pellet at room temperature (RT) for ~5 min to evaporate residual methanol (*see* **Note 13**).

3.1.2 Chloroform/Methanol Precipitation

1. Add 5× the sample volume of ice-cold chloroform/methanol.
2. Vortex vigorously for 15 s.
3. Centrifuge at 20,000× *g* for 1 h at 4 °C.
4. Carefully aspirate the liquid phases without disturbing the protein disc (*see* **Note 14**).
5. Add 1 mL of ice-cold methanol and vortex to detach the protein disc from the wall of the microtube.
6. Centrifuge the sample at 20,000× *g* for 15 min at 4 °C.
7. Gently aspirate the supernatant and air dry the protein pellet at RT for ~5 min to evaporate residual methanol (*see* **Note 13**).

3.1.3 Trichloroacetic Acid (TCA) Precipitation

1. Add ice-cold 100% TCA in a 1 to 8 ratio (v/v) to the sample.
2. Vortex vigorously for 15 s.
3. Incubate for 2 h at 4 °C.
4. Centrifuge at 20,000× *g* for 30 min at 4 °C (*see* **Note 12**).
5. Gently aspirate the supernatant.
6. Add 1 mL of ice-cold acetone to the pellet.
7. Briefly vortex to detach the pellet from the bottom of the tube and incubate for 1 h at −20 °C.
8. Centrifuge at 20,000× *g* for 30 min at 4 °C.
9. Wash the pellet twice with 1 mL of ice-cold acetone, each, and repeat **step 8**.
10. Gently aspirate the supernatant and air dry the protein pellet at RT for ~5 min to evaporate residual acetone (*see* **Note 13**).

3.2 In-Solution Protein Digestion

Independent of the precipitation method, protein samples can be digested in solution using either RapiGest or urea. While RapiGest-resuspended samples can be stored in the freezer, this is not the case for urea. Freezing of samples containing high urea concentrations results in urea precipitation, complicating further processing after thawing, and can lead to unwanted side reactions such as the modification of amino acids or protein/peptide N-termini [16].

3.2.1 Urea Digestion

1. Resuspend the protein pellet in 8 M urea 0.1 M TEAB solution (*see* **Note 15**).
2. Incubate the sample at 600 rpm for 45 min at 37 °C in a thermomixer.
3. Reduce protein disulfide bonds using DTT (5 mM final concentration) at 600 rpm for 30 min at 56 °C in a thermomixer.
4. Alkylate free SH groups using acrylamide or chloroacetamide at a final concentration of 20 mM or 10 mM, respectively, at 600 rpm for 30 min at RT in a thermomixer (in the case of chloroacetamide incubate in the dark).
5. Quench the alkylation reaction by adding additional DTT (5 mM final concentration) to the sample and incubate at 600 rpm for 10 min at RT in a thermomixer.
6. Dilute the sample to 4 M urea (final concentration) using 0.1 M TEAB.
7. Add rLysC at an enzyme-to-protein ratio of 1:100 and incubate at 37 °C for 8 h (*see* **Note 16**).
8. Dilute the sample to 1.3 M urea (final concentration) using 0.1 M TEAB.
9. Add trypsin at an enzyme-to-protein ratio of 1:100 and incubate at 37 °C overnight.
10. Acidify the sample by addition of either FA, AA, or TFA (0.5% final concentration) to stop enzymatic activity.
11. Desalt the sample using Oasis cartriges (Subheading 3.4.1, *see* **Note 17**).

3.2.2 RapiGest Digestion

1. Resuspend the protein pellet in 0.2–0.5% RapiGest, 0.1 M ABC. Adjust the volume to the amount of protein (*see* **Note 18**).
2. Heat the sample for 5 min at 95 °C to dissolve the pellet.
3. Briefly spin down the sample to collect possible liquid precipitate in the lid of the microtube.
4. Sonicate the sample using a water bath sonicator for 5 min to help the pellet to dissolve. If the pellet does not dissolve, (*see* **Note 19**).
5. Reduce protein disulfide bonds using DTT (5 mM final concentration) at 600 rpm for 30 min at 56 °C in a thermomixer.
6. Alkylate free SH groups using acrylamide or chloroacetamide at a final concentration of 20 mM or 10 mM, respectively, at 600 rpm for 30 min at RT in a thermomixer (in the case of chloroacetamide incubate in the dark).
7. Quench the alkylation reaction by adding additional DTT (5 mM final concentration) to the sample and incubate at 600 rpm for 10 min at RT in a thermomixer.

8. Add trypsin at an enzyme-to-protein ratio of 1:100, dilute the sample to a final concentration of 0.1% RapiGest with 0.1 M ABC, and incubate overnight at 37 °C.
9. Acidify the sample by the addition of TFA (final concentration of 1%) and incubate at 800 rpm for 30 min at 37 °C in a thermomixer to hydrolyze RapiGest.
10. Centrifuge at 20,000× *g* for 10 min at RT.
11. Carefully transfer the supernatant without disturbing the pellet (hydrolyzed RapiGest) to a fresh tube (*see* **Note 20**).
12. The sample can be desalted using Oasis cartridges (Subheading 3.4.1, *see* **Note 21**).

3.3 On-Bead Protein Digestion

3.3.1 Single-Pot, Solid-Phase-Enhanced Sample Preparation (SP3) Digestion

This protocol is especially useful when only small protein amounts are available as it does not require a prior protein precipitation step and most non-MS compatible chemicals are depleted. Any kind of lysosome-enriched samples can be prepared for MS-based proteomics with this approach. The protocol does need, however, specialized equipment (a PCR tube-compatible magnetic stand with a sufficiently strong magnet) and larger amounts of protein cannnot be efficiently processed with the setup described here.

1. Transfer the desired protein amount (10–50 μg) into a fresh PCR tube and bring the volume to 50 μL by addition of 0.5 M TEAB.
2. Reduce protein disulfide bonds using DTT (5 mM final concentration) at 600 rpm for 30 min at 56 °C in a thermomixer.
3. Alkylate free SH groups using acrylamide or chloroacetamide at a final concentration of 20 mM or 10 mM, respectively, at 600 rpm for 30 min at RT in a thermomixer (in the case of chloroacetamide incubate in the dark).
4. Quench the alkylation reaction by adding additional DTT (5 mM final concentration) to the sample and incubate for 10 min at RT.
5. Transfer the sample into a fresh PCR tube containing the pre-mixed SP3 beads (*see* **Note 7**) and mix by gently pipetting.
6. Add 3x the sample volume of 96% ethanol to the protein/bead mixture and incubate in a thermomixer at 1000 rpm for 5 min at 24 °C (*see* **Note 22**).
7. Place the tubes in the magnetic rack and carefully aspirate the liquid phase without disturbing the beads.
8. Wash the beads with 180 μL of 80% ethanol by gently pipetting up and down (*see* **Note 23**).
9. Repeat **steps** 7 and **8** two more times (*see* **Note 24**).

10. Resuspend the beads in 0.5 M TEAB (*see* **Note 25**), add rLysC at an enzyme to protein ratio of 1:100 (w/w), and incubate at 1000 rpm for 2 h at 37 °C.
11. Add trypsin at an enzyme-to-protein ratio of 1:100 (w/s) and incubate at 1000 rpm overnight at 37 °C.
12. Place the tubes in the magnetic rack and transfer the supernatant containing the peptide mixture into a fresh microtube.
13. Acidify the digestion solution to stop tryptic activity by adding of either AA, FA, or TFA (0.5% final concentration).
14. Samples can be vacuum-dried and later desalted with STAGE-tips (see Subheading 3.4.2, *see* **Note 21**).

3.4 Peptide Desalting

For the desalting of peptides, typically reversed-phase stationary phases are used. The choice of which format, and hence capacity, depends on the sample's peptide amount. For larger amounts, originating from urea or RapiGest digestion, typically cartridge-based solutions are used, such as Oasis or Sep-Pak. Oasis cartridges are available in different sizes. For lysosome-enriched samples, typically 1 cc/10 mg cartridges are sufficient. For small amounts of peptides derived from SP3 digests, STAGE tips provide a sufficient capacity. They are also more convenient in handling and result in reduced peptide loss.

3.4.1 Oasis Solid Phase Extraction (SPE) Cartridges

1. For all steps except for peptide binding and peptide elution, the SPE vacuum manifold in combination with a vacuum pump is used as it enables the parallel processing of high sample numbers. In case only a few samples are processed, a syringe in combination with an adaptor can be used (*see* **Note 8**).
2. Activate the cartridge with 4 mL of Elution Buffer.
3. Wash the cartridge with 4 mL of Washing Buffer.
4. Place a fresh tube below the cartridge to collect the flow-through fraction and load the peptide mixture into the cartridge using the Sep-Pak reservoir adapter and a syringe (*see* **Note 26**).
5. Place the cartridge back onto the SPE vacuum manifold holder and wash it with 10 mL of Washing Buffer. The wash fraction can be discarded.
6. Place a fresh tube below the cartridge to collect the eluate fraction and load 500 μL of Elution Buffer. Apply pressure with the help of the reservoir adapter and the syringe.
7. Dry the sample using a vacuum centrifuge.
8. The dried samples can either be stored at −20 °C, or resuspended in Resuspension Buffer 1 (compatible with peptide assay kit) for assessing the peptide content using a fluorometric peptide assay, or dissolved in Resuspension Buffer 2 for MS analysis (*see* **Note 27**).

3.4.2 Stop and Go Extraction (STAGE) Tips

1. Assemble the STAGE tip and place it into a tip holder in a 2-mL microtube (*see* **Note 28**).
2. All steps are performed by centrifugation at 2500× *g* for 1 min at RT.
3. Activate the C_{18} material by loading the tip once with 20 μL of 100% methanol, twice with 20 μL of Elution Buffer, and twice with 20 μL of Washing Buffer (*see* **Note 29**).
4. Place the tip/holder into a fresh 2-mL microtube to collect the flow-through/wash, load the peptide mixture (*see* **Note 29**), and subsequently wash the tip with 100 μL of Washing Buffer.
5. Place the tip/holder into a fresh 1.5-mL microtube to collect the eluate fraction and elute the peptides twice with 20 μL of Elution Buffer.
6. Dry the combined eluate fractions with a vacuum centrifuge. Dried peptides can be stored at −20 °C or resuspended in Resuspension Buffer 2 for MS analysis (*see* **Note 30**).

3.5 LC-MS/MS Analysis

Loading similar amounts of sample, either based on the results of a peptide assay (typically 200–400 ng) or by taking a fixed fraction of the sample (in the case of SP3, typically 20% for 15 μg protein input), increases reproducibility and will yield better results for label-free quantification. Depending on the type of chromatographic system and mass spectrometer, parameters vary. The settings described here deliver good results with an Orbitrap Fusion Lumos. Samples are loaded onto the chromatographic column with 850 nL/min Buffer A for 16 min, followed by elution of the peptides with a linear gradient from 5% to 35% of Buffer B at a flow rate of 300 nL/min. To achieve maximum coverage of lysosome-enriched fractions, a gradient length of 120 min can be used, but also shorter gradients result in good coverage of the lysosomal proteome [17]. For DIA acquisition, the following MS settings are used: MS1 scans are acquired at a mass range of *m/z* 350–1200 in the Orbitrap analyzer at a resolution of 120,000, a normalized HCD collision energy of 30%, a maximum injection time of 50 ms, an AGC target setting of 1×10^6, and a cycle time of 2 s. For MS2 scans, a total of 36 static windows (m/z 24.1 each) are defined, which cover the m/z 350–1200 mass window with an overlap of m/z 0.5. MS2 scans are acquired from *m/z* 200–2000 at a resolution of 30,000 in the Orbitrap analyzer with a maximum injection time of 54 ms, an AGC target setting of 5×10^5, and a normalized HCD collision energy of 30%. A typical total ion chromatogram obtained from a DIA analysis of a lysosome-enriched sample is shown in Fig. 2. Obtained *.raw files are analyzed using Spectronaut for simultaneous identification of peptides/proteins and label-free quantification between different

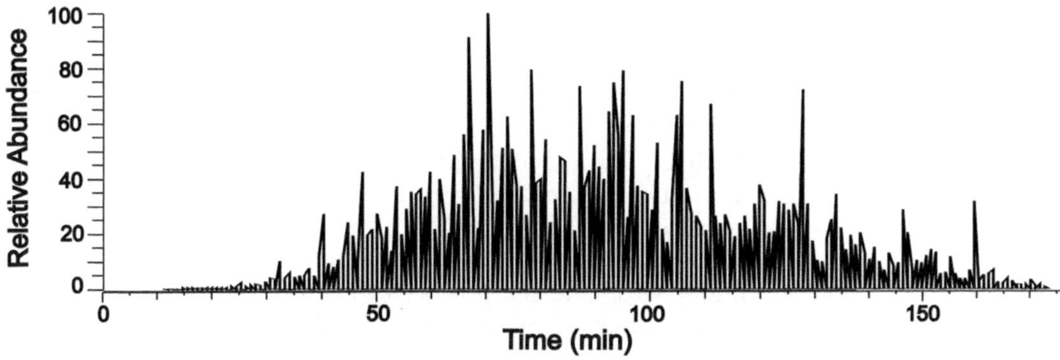

Fig. 2 Typical total ion chromatogram (TIC) obtained from the data-independent acquisition (DIA) liquid chromatography-tandem mass spectrometry (LC-MS/MS) analysis of 250 ng peptides obtained from the tryptic digestion of a lysosome-enriched sample from HEK293 cells. Enrichment was performed using superparamagnetic iron oxide nanoparticles (SPIONs), the sample was digested with single-pot, solid-phase-enhanced sample-preparation (SP3), and peptides were desalted by stop and go extraction (STAGE) tips. DIA analysis was performed on an Orbitrap Fusion Lumos mass spectrometer using a 120 min gradient. (Raw data are provided from [23])

conditions. Based on the type of cells, the appropriate *. fasta file has to be selected (https://www.uniprot.org/). With respect to data processing, the standard settings employed for proteomic analysis can be applied; no special considerations are necessary for lysosome-enriched fractions. Typical parameters are as follows: fragment ions for each peptide: 3–6; enzyme: trypsin/P; number of allowed miss cleavage sites: 2; fixed modification: propionamide or carbamidomethyl (dependent on the alkylation reagent) at cysteine; variable modifications: acetylation at protein N-termini and oxidation of methionine. Peptide and protein identification are performed at a false discovery rate (FDR) of 1%. For peptide/protein quantification, a q-value cutoff of 0.01 and 0.05, respectively, can be applied, and missing value imputation is disabled. Aside from Spectronaut, several other software packages are available, which can be utilized for the processing of DIA data (reviewed in [18]). Independent of the enrichment and digestion approach, such analyses typically identify >5000 proteins, of which only a subset is to date verified as truly lysosomal [1, 2, 17].

4 Notes

1. Caution must be taken while preparing the stock solution. TCA powder is highly acidic and can cause skin burns and eye damaging. The powder must be handled wearing protective equipment and gloves.
2. Urea powder should be prepared fresh on the day of the experiment and not be reused or frozen.

3. TEAB should be used to dissolve urea in order to decrease the degree of carbamylation at N-termini and Lys-residues [19]. Furthermore, the solution should be prepared as fresh as possible.
4. Both DTT and acrylamide are hygroscopic powders stored at 4 °C. Before preparing stock solutions, the containers should be allowed to warm to RT in a desiccator. These chemicals should further be handled only in a fume hood as they are highly toxic and the powder can easily be inhaled. After generating stock solutions, single-use aliquots should be prepared and stored at −20 °C to avoid freeze-thaw cycles, which could affect reactivity. Chloroacetamide is light sensitive when in solution; therefore, prepare the working solution fresh and incubate in the dark.
5. Both proteolytic enzymes must be dissolved and aliquoted in single-use tubes and stored at −20 °C to avoid repeated freeze-thaw cycles.
6. Prepare 1–10% (w/v) RapiGest stock solution (depending on the amount purchased) in MS water or 0.1 M ABC and store as aliquots at −20 °C.
7. The two types of beads are mixed in a 1:1 ratio and subsequently combined with the sample in a 20:1 ratio (w/w, combined beads relative to protein amount in sample). For example, in order to process a sample containing 30 μg of protein, 300 μg of each type of SP3 bead should be used, amounting to a total of 600 μg of beads [15]. Prior to incubation with the samples, the bead mixture should be washed three times with 0.5 M TEAB buffer to remove residual storage solution. Before combination with the sample, the beads should be resuspended in water to obtain a 10 μg beads/μL solution.
8. The system is operated with a vacuum pump which has to be purchased separately.
9. The capacity of one C_{18} disk is 3–5 μg of peptides, and a combination of several disks allows for an increase in the binding capacity of the STAGE tip. Do not stack more than ten C_{18} disks into one tip, as they might shift in the tip during centrifugation and sample flow rates decrease with increasing disk numbers.
10. In the case of highly diluted lysosome-enriched samples, it is possible to pellet down the lysosomes at 20,000× *g* for 30 min at 4 °C, but only when the enrichment strategy does not include elution buffers containing detergents. The resulting pellet can be dissolved in urea/RapiGest-containing buffers, and it can be proceeded with in-solution digestion as described

in Subheading 3.2 without the need for an additional protein precipitation step.

11. This step can also be performed overnight at −20 °C.
12. After centrifugation, the protein pellet should be visible at the bottom of the tube.
13. The protein pellet can be stored at this point at −80 °C until further use. However, storage often results in reduced solubilization efficiency.
14. After centrifugation, the protein pellet should be visible as a disk at the interface between the chloroform solution (bottom) and the aqueous methanol solution (top). In the case of insufficient amounts of protein, patches of precipitate rather than a disk are typically visible. Aspirate the aqueous solution above the disk and tilt the tube so the protein disc locates to the wall of the tube. Now, the bottom liquid can be aspirated as well. Ideally, this step is performed with a 0.5 or 1-mL Hamilton glass syringe.
15. The sample should be resuspended in a volume of urea buffer to aim for a protein concentration of 4–8 μg/μL. This step can require more time because the protein pellet can be hard to dissolve, especially for acetone and TCA precipitates. Volumes of <30 μL should be avoided as they complicate handling and therefore influence reproducibility.
16. rLysC also cleaves lysine residues, similar to trypsin. However, rLysC is active at higher urea concentrations, enabling digestion under stronger denaturing conditions. Therefore, when added prior to trypsin, it enhances protein digestion, resulting in lower missed cleavage rates [20].
17. Samples cannot be frozen at this point due to solubility issues and potential side reactions with urea. It is advised to directly proceed with peptide desalting.
18. Maintain a final ratio of 0.1% RapiGest/1 μg of protein. For example, for 100 μg of protein, use 50 μL of 0.2% RapiGest or 20 μL of 0.5% RapiGest.
19. If the pellet insufficiently dissolves, add trypsin at an enzyme to protein ratio of 1:500 (w/w) and incubate at 400 rpm for 30 min at 37 °C in a thermomixer.
20. After the incubation, samples should turn cloudy, and often white particles are visible. In case RapiGest does not precipitate properly, the protein concentration is lower than anticipated. Potential strategies to affect RapiGest precipitation can be found here [21]. Additionally, it is possible to incubate the sample for 1 h at 4 °C to enhance RapiGest precipitation.

21. The protocol could be temporarily paused at this point prior to desalting. However, it is not optimal to keep the sample in a TFA-containing solution for prolonged times.
22. If the sample/beads/ethanol mixture does exceed the PCR tube's volume, it is advisable to perform the binding step in a 1.5-mL microtube and to transfer the beads after incubation to a PCR tube for the following steps (or to use a magnetic stand compatible with 1.5 mL tubes). For incubation of beads and samples, longer timing does not enhance the amount of proteins binding to the beads [15].
23. To remove the wash fraction, place the tube into the magnetic rack and wait until all the beads are attached to the wall. During the washing steps, beads tend to clump. This indicates that proteins have bound to the beads. Try to resuspend the beads thoroughly to ensure removal of MS-incompatible chemicals.
24. If high concentrations of detergent are present in your lysis buffer (e.g., more than 1% of SDS or Triton X-100), perform additional washing steps.
25. Adjust the digestion volume to the amount of protein input. For example, 50–100 μL of 0.5 M TEAB can be used for samples containing 30 μg. Make sure that the beads are completely covered by the buffer.
26. Never load less than 500 μL on the column, and ensure that the sample has an acidic pH to allow for optimal binding of peptides to the reversed phase material. If peptides are resuspended in a lower volume, dilute to 500 μL with Washing Buffer. It is advisable to store the flow-through fraction until samples are analyzed by LC-MS/MS, in case peptide binding to the stationary phase did not work properly.
27. Performing the estimation of peptides obtained after digestion and desalting helps to load an equal amount for LC-MS/MS measurements, increasing the reproducibility. After the peptide assay, the sample either has to be dried again and resuspended in Resuspension Buffer 2 or FA should be added to the sample to a final concentration of 5% (if the peptide concentration is sufficiently high).
28. The desired number of C_{18} disks is cut and stacked into a 200-μL pipette tip with the help of a blunt-tipped needle. Instructions for STAGE tip preparation can be found in [22]. Adjust the number of C_{18} disks placed into each pipette tip depending on the protein amount. One C_{18} disk can bind 3–5 μg of peptides. Also, pre-packed desalting tips are commercially available.
29. After the washing steps have been performed, check if any of the disks have moved or become separated. If this is the case, it

is recommended to replace the tip with a new one, as otherwise gaps/channels may form through which the sample bypasses the C_{18} material, resulting in strongly reduced binding behavior. If the sample is not passed completely after 1 min of centrifugation, increase the centrifugation time to up to 2 min (or more) rather than increasing the centrifugation speed and repeat, if needed. Proceed only when the sample has completely passed through the tip. Proceed with all steps in one experimental sequence to prevent drying of the C_{18} disks.

30. For SP3 digests, peptide assays often do not result in good readings due to insufficient sensitivity.

References

1. Akter F, Bonini S, Ponnaiyan S et al (2023) Multi-cell line analysis of lysosomal proteomes reveals unique features and novel lysosomal proteins. Mol Cell Proteomics 22(3):100509
2. Muthukottiappan P, Winter D (2021) A proteomic view on lysosomes. Mol Omics 17(6): 842–859
3. Chapel A, Kieffer-Jaquinod S, Sagné S et al (2013) An extended proteome map of the lysosomal membrane reveals novel potential transporters. Mol Cell Proteomics 12(6): 1572–1588
4. Di Lorenzo G, Velho RV, Winter D et al (2018) Lysosomal proteome and secretome analysis identifies missorted enzymes and their nondegraded substrates in mucolipidosis III mouse cells. Mol Cell Proteomics 17(8): 1612–1626
5. Wyant GA, Abu-Remaileh M, Frenkel EM et al (2018) NUFIP1 is a ribosome receptor for starvation-induced ribophagy. Science 360(6390):751
6. Zhang YY, Fonslow BR, Shan B et al (2013) Protein analysis by shotgun/bottom-up proteomics. Chem Rev 113(4):2343–2394
7. Kulevich SE, Frey BL, Kreitinger G et al (2010) Alkylating tryptic peptides to enhance electrospray ionization mass spectrometry analysis. Anal Chem 82(24):10135–10142
8. Dupree EJ, Jayathirtha M, Yorkey H et al (2020) A critical review of bottom-up proteomics: the good, the bad, and the future of this field. Proteomes 8(3):14
9. Itzhak DN, Tyanova S, Cox J et al (2016) Global, quantitative and dynamic mapping of protein subcellular localization. elife 5:5
10. Arora D, Hackenberg Y, Li J et al (2023) Updates on the study of lysosomal protein dynamics: possibilities for the clinic. Expert Rev Proteomics 20:47
11. Mosen P, Sanner A, Singh J et al (2021) Targeted quantification of the lysosomal proteome in complex samples. Proteomes 9(1):4
12. Graham JM (2001) Isolation of lysosomes from tissues and cells by differential and density gradient centrifugation. Curr Protoc Cell Biol 4:3.4.1–3.4.22
13. Walker MW, Lloyd-Evans E (2015) A rapid method for the preparation of ultrapure, functional lysosomes using functionalized superparamagnetic iron oxide nanoparticles. In: Platt F, Platt N (eds) Lysosomes and lysosomal diseases, pp 21–43
14. Müller T, Winter D (2017) Systematic evaluation of protein reduction and alkylation reveals massive unspecific side effects by iodine-containing reagents. Mol Cell Proteomics 16(7):1173–1187
15. Hughes CS, Moggridge S, Müller T et al (2019) Single-pot, solid-phase-enhanced sample preparation for proteomics experiments. Nat Protoc 14(1):68
16. Kollipara L, Zahedi RP (2013) Protein carbamylation: in vivo modification or in vitro artefact? Proteomics 13(6):941–944
17. Ponnaiyan S, Akter F, Singh J et al (2020) Comprehensive draft of the mouse embryonic fibroblast lysosomal proteome by mass spectrometry based proteomics. Sci Data 7(1):68
18. Zhang FF, Ge W, Huang L et al (2023) A comparative analysis of data analysis tools for data-independent acquisition mass spectrometry. Mol Cell Proteomics 22(9):100623
19. Sun SS, Zhou JY, Yang W et al (2014) Inhibition of protein carbamylation in urea solution using ammonium-containing buffers. Anal Biochem 446:76–81
20. Li Q, Feng Y, Tan MJ et al (2017) Evaluation of endoproteinase Lys- C/trypsin sequential

digestion used in proteomics sample preparation. Chin J Anal Chem 45(3):316–321

21. Mosen PR, Hardt R, Winter D (2021) RapiGest precipitation depends on peptide concentration. Proteomics 21(20):e2100129
22. Rappsilber J, Ishihama Y, Mann M (2003) Stop and go extraction tips for matrix-assisted laser desorption/ionization, nanoelectrospray, and LC/MS sample pretreatment in proteomics. Anal Chem 75(3):663–670
23. Bonini S, Winter D (2024) Two-step enrichment facilitates background reduction for proteomic analysis of lysosomes. J Proteome Res 23(8):3393–3403

Chapter 10

Analyzing Endolysosome–Mitochondria Membrane Contact Sites by Thin Section TEM of Human Fibroblasts

Claudia Parra-Ruiz, Carlos Enrich, and Silvana Zanlungo

Abstract

Membrane contact sites (MCS) are dynamic nanoregions of close apposition between two different organelles, functioning as discrete lipid or ion transfer sites. This new concept in cell biology involves unique proteins at both membrane sites, named tethers, and emerges in early observations by transmission electron microscopy (TEM). Currently, this technique still constitutes a valuable tool for MCS visualization and quantification. In the last decade, Lysosomal Storage Diseases (LSD) have been instrumental in studying the MCS between lysosomes (Ly), or endolysosomes (EL), and other organelles in close proximity such as mitochondria or the endoplasmic reticulum (ER). At present, the analysis of composition, functioning, and alterations/rewiring of MCS in health and disease represents an innovative area of research for designing therapeutic strategies in a variety of pathologies. Here, we describe procedures for chemical fixation using the Flat Embedding technique to characterize and quantify the MCS between LE/Lys and mitochondria in human fibroblasts by thin-section TEM.

Key words Lysosomes, Mitochondria, Membrane contact sites, Transmission electron microscopy-chemical fixation, Image analysis, Lysosomal storage diseases, Flat embedding

1 Introduction

Early electron microscopy studies described close proximities between different organelles [1]. Later, researchers investigated whether these intracellular contacts between organelles provided a functional advantage and were necessary for cell homeostasis. It took even more time to establish membrane contact sites (MCSs) as a new paradigm in cell biology for intracellular communication [2]. It is now widely accepted that apposition via MCS is directly related to exchange and signaling [3–7]. Therefore, compartments in the cell do not function as isolated entities but as a dynamic and regulated ensemble that facilitates the transport of lipids and ions. The formation and functionality of MCS require specific proteins and lipids that serve as tethers. Some of these proteins provide additional functions beyond tethering and traffic of metabolites,

Stephanie M. Hughes and Indranil Basak (eds.), *Lysosomes: Methods and Protocols*, Methods in Molecular Biology, vol. 2976, https://doi.org/10.1007/978-1-0716-4844-5_10,

including organelle trafficking, endosome maturation and positioning, membrane dynamics, Ca^{2+} signaling, and organelle architecture fusion and fission [8–11].

MCS between mitochondria and endolysosomes (LE/Lys) has recently been valued as a relevant crosstalk for the regulation of cellular homeostasis, where both organelles fulfill essential functions, and dysregulation of LE/Lys-mitochondria interaction has been identified as an important pathogenic mechanism in several diseases [3, 12]. Nevertheless, the study of these contacts remains challenging, as the structural heterogeneity of the endolysosomal compartment, i.e., the distinction between late endosomes, multivesicular bodies (MVBs), pre-lysosomes, lysosomes, or autolysosomes, is not trivial. Indeed, TEM remains the best alternative for precisely distinguishing between these structural entities. Using advanced imaging techniques, multiple reports demonstrated MCS between mitochondria and lysosomes in different cell types under healthy conditions [13–15], which has been shown to be independent of autophagosome biogenesis or mitophagy [16, 17] and allow bidirectional regulation of both organelles. Mitochondria heavily rely on contact sites to transfer metabolites, including ions and lipids, as they are largely excluded from vesicular trafficking [7, 18]. Therefore, MCSs play a crucial role in maintaining metabolic homeostasis. This idea is supported by recent research showing that alterations in the formation and regulation of mitochondria-LE/Lys MCS, or modifications involving rewiring with other organelles, can be relevant to understanding the progression of various human pathologies [3, 19]. In previous studies, we and others have demonstrated that changes in mitochondrial cholesterol levels are detrimental and negatively impact mitochondrial functionality [20–23]. Accordingly, a decrease in lysosome-ER contact sites and an increase in lysosome–mitochondria contact sites have been demonstrated in Niemann–Pick type C1 mutant cells [21, 24, 25], produced by deficiency or malfunctioning in the lysosomal protein NPC1 and the subsequent cholesterol accumulation in lysosomes. Interestingly, in iPSC-derived neurons harboring a *GBA1* mutation, as well as a pharmacological model of Gaucher Lysosomal Storage Disease (LSD) induced by conduritol-β-epoxy (CβE) treatment, sustained contact sites between lysosomes and mitochondria were described [26].

On the other hand, mitochondria-LE/Lys MCS play a crucial role in transferring iron (Fe^{2+}) between organelles. Specifically, LE/Lys transfers Fe^{2+} from endosomes containing transferrin receptors to mitochondria, which is necessary for the biogenesis of heme and iron–sulfur clusters [27]. Additionally, mitochondria-LE/Lys MCS modulate intracellular Ca^{2+} dynamics. A recent study by Peng et al. [28] demonstrated that the lysosomal channel TRPML mediates Ca^{2+} efflux to mitochondria, preferentially through MCS, and shows that mitochondrial Ca^{2+} dynamic is modified in Mucolipidosis type IV, resulting from the loss of TRPML1 function. This provides evidence of convergent dysfunction of both organelles in the disease [28].

Two methodologies are available for analyzing LE/Lys-mitochondria MCS in human fibroblasts using TEM. The first method involves fixing cells cultured as monolayers in their culture dish, followed by gentle cell scraping and pelleting within microcentrifuge tubes. The second technique, flat embedding, involves growing cells on glass coverslips, fixing and subsequently embedding in place, to avoid the centrifugation steps and thus reducing the likelihood of any morphological alterations. Even though both techniques yield satisfactory results, flat embedding is better at preserving structures, especially the mitochondrial network and cytoskeleton arrangement. Accordingly, this chapter describes procedures for analyzing LE/Lys-mitochondria MCS extent in human fibroblasts using TEM from chemical-fixed samples and processed using flat embedding methodology. In the Materials and Methodology sections, we present the information divided into two main modules: first, concerning the generation of thin TEM sections starting from cell culture, and second, image capture and analysis, including quantification of LE/Lys-mitochondria MCS extent.

2 Materials

Analytical-grade reagents and double-distilled or ion-exchanged water must be used to prepare all solutions. Some of the reagents used in TEM sample preparation are toxic, and therefore, all preparations must be carried out under a chemical hood. Follow all safety and waste disposal regulations outlined in the product technical specifications and institutional procedures.

2.1 Chemical Fixation and Sample Preparation

2.1.1 Cell Culture

1. This study used Coriell Institute Human Skin Fibroblast (HSF) GM05659, although the methodology presented can be applied to other cell cultures.
2. 12-mm glass coverslips.
3. 6 multiwell cell culture plates.

2.1.2 Fixation Buffer: 0.2 M Phosphate Buffer (PB) pH 7.4

1. Solution A: dissolve 2.76 g of sodium phosphate monobasic monohydrate ($NaH_2PO_4{\cdot}H_2O$) in 80 mL of Milli-Q water, stir with a magnetic stirrer, and make up to 100 mL. Store at 4 °C. The solution expires approximately after 30 days. It is advisable to filter with a 0.22-μm filter before final mixing with solution B.
2. Solution B: dissolve 17.80 g of sodium dihydrogenphosphate dihydrate ($NaH_2PO_4{\cdot}2H_2O$) in 400 mL of MilliQ water, stir with a magnetic stirrer, and make up to 500 mL. Store in an

amber bottle at room temperature; it expires approximately after 30 days. It is advisable to filter with a 0.22-μm filter before final mixing with solution A.

3. To obtain 100 mL of 0.2 M PB buffer at pH 7.4, mix solutions A and B in the following proportion: 19.6 mL solution A + 80.4 mL solution B. The final mixture is stored at 4 °C, and it expires approximately after 15 days (ensure it is not contaminated before reuse).

2.1.3 Fixative Solution: 3% Glutaraldehyde in 0.1 M PB pH 7.4

To prepare 10 mL of fixative solution: mix 5 mL of PB 0.2 M, 1.2 mL of glutaraldehyde 25% solution, and 3.8 mL of double-distilled water. Store at 4 °C. Handling glutaraldehyde and fixative solution must always be done under a fume hood and with gloves.

2.1.4 Post-fixation Solution: 1% Osmium Tetroxide–0.8% Potassium Ferrocyanide in 0.1 M PB buffer pH 7.4

1. Mix 2.5 mL of 4% osmic acid anhydride and 2.5 mL of 0.1 M PB pH 7.4. Protect the solution against light with aluminum foil (*see* **Note 1**).
2. Weigh 150 mg of potassium ferrocyanide and dissolve it in 5 mL of 0.1 M PB pH 7.4. Protect the solution against light with an aluminum foil.
3. To obtain 10 mL of post-fixation solution, mix solutions from **steps 1** and **2** just before use.

2.1.5 Dehydration

Ethanol in water (50, 70, 90, 96, and 100%). Store solutions at 4 ° C.

2.1.6 Epoxy Resin (Pure Spurr) Preparation

1. NSA: Nonenyl Succinic Anhydride.
2. ERL: Cycloaliphatic Epoxide Resin ERL4221.
3. DER: Dow Epoxy Resin—Diglycidyl Ether of Polypropylene Glycol.
4. DBP: Dibutyl Phthalate.
5. DMAE S1: 2-dimethylaminoethanol S1.

2.1.7 Embedding and Resin Polymerization

1. BEEM embedding capsules, Size 00, 8 mm I.D.
2. Laboratory oven.

2.1.8 Sectioning

1. Ultramicrotome with a diamond knife (unless the researcher has experience in sectioning, it is highly recommended to count on specialized facilities to obtain the sections of embedded samples).
2. Copper grids 200 Mesh.

2.1.9 Contrast of Ultrathin Sections

1. UranyLess EM Stain.
2. Lead Citrate 3% (ready-to-use solution).
3. Sodium Hydroxide pellets.

2.1.10 Other Materials and Reagents Necessary for TEM Sample Preparation

1. High-precision tweezers for grid and coverslip manipulation.
2. 10 cm Petri dishes.
3. Single-use Pasteur pipettes.
4. Plastic container.

2.2 Microscope and Image Analysis

1. A Transmission Electron Microscopy JEOL JEM-1010, operating at 80 kV and fitted with a Gatan Orius SC1000 (model 832) digital camera, was used in our studies.
2. Software ImageJ [29] for image processing and GraphPad Prism for statistical data analysis.

3 Methods

The methodology is divided into two main sections: sample handling (which includes grid preparation) and image acquisition and analysis.

3.1 Sample Handling and Grid Preparation: Chemical Fixation and Sample Processing Using Flat Embedding Method

The flat embedding technique enables the acquisition of resin blocks from monolayers of cells on coverslips. This technique is used to process samples. This technique is advantageous for maintaining whole cell morphology, including inner structures, a key issue for membrane contact studies. The fixation (Subheading 3.1.1) to Resin Infiltration (Subheading 3.1.4) steps can be performed with cells in coverslips without removing them from the multiwell plate. In each step, the solution volume necessary for a single well with a coverslip depends on the size of the plate well, being the criterion that the coverslips should be well-covered. The wells of the plate must be large enough to allow the manipulation of the coverslips and their extraction.

3.1.1 Cell Culture and Fixation

1. Culture the cells on glass coverslips located on a multiwell plate until the culture is at the appropriate stage for investigation (confluence 50–80%). Once confluency is reached, remove the culture medium (without allowing the sample to dry), wash with PB solution, and gently add the fixative solution. Incubate for 1 h at room temperature (*see* **Note 2**).
2. Continue with the following processing steps or seal the plate with parafilm and keep the samples immersed in the fixative buffer at 4 °C for long periods (*see* **Note 3**).

3.1.2 Post-fixation

Perform all the steps, maintaining the coverslips on the plate.

1. After fixation, wash the coverslips three times (5 min each) with PB at room temperature to remove the excess fixative.
2. Prepare the post-fixation solution just before use.

3. Add the osmium post-fixation solution to the sample and leave for 30–45 min at 4 °C in the dark (inside the osmium storage refrigerator).
4. Rinse the coverslips three times (5 min each) with double-distilled water at room temperature to remove osmium excess. The next step of dehydration continues the same day.

3.1.3 Dehydration

Maintaining the coverslips on the plate, dehydrate with successive changes of increasing ethanol solutions (1–2 mL) at room temperature.

1. Incubate coverslips for 10 min in 50% ethanol.
2. Incubate coverslips for 10 min in 70% ethanol.
3. Incubate coverslips for 10 min in 80% ethanol.
4. Incubate coverslips for 10 min in 90% ethanol (three changes).
5. Incubate coverslips for 10 min in 96% ethanol (three changes).
6. Incubate coverslips for 10 min in 100% ethanol (three changes). On the same day, the infiltration with the resin step continues.

3.1.4 Resin Preparation and Sample Infiltration

1. For resin (pure Spurr) generation, weigh the following components in order using plastic Pasteur pipettes and put them into a plastic container: 26 g NSA, 10 g ERL, 6 g DER, 0.8 g DBP, and 0.4 g DMAE S1 (*see* **Note 4**).
2. Mix with a magnetic stirrer for about 15 min. Leave to rest for at least 30 min before using it so that the bubbles generated during agitation disappear.
3. Store pure Spurr resin at room temperature and in the fume hood during the embedding process. For storage and conservation: this resin already has the catalyst (DMAE), so keep it for a maximum of 48 h at room temperature and then keep the product at 4 °C until the next day. At this temperature, it has a duration of 1 week.
4. Once the resin has been prepared, make a Spurr resin battery mixed with ethanol 100% of increasing concentration. This is for sample incubation, which can be performed at room temperature and does not need to be shaken.
5. Incubate coverslips for 1 h in one volume of pure Spurr/three volumes of ethanol.
6. Incubate coverslips for 1 h in two volumes of pure Spurr/two volumes of ethanol.
7. Incubate coverslips for 1 h in three volumes of pure Spurr/one volume of ethanol.
8. Incubate coverslips overnight in pure Spurr.

9. Next day, incubate coverslips for 6 h in fresh pure Spurr. Polymerization steps continue on the same day. Until this, keep the samples in pure Spurr.

3.1.5 Embedding and Resin Polymerization

1. As a preliminary step and to prevent the appearance of bubbles that can get stuck interfering with cells during the polymerization process, we recommend that 2 or 3 days before cell culture processing (Subheadings 3.1.1, 3.1.2, 3.1.3 and 3.1.4), add pure Spurr resin inside BEEM capsules almost to the top, that is, 1–2 mm below the rim. Fill as many BEEM capsules as coverslips and include a couple of extra ones just in case. Let polymerize in a 60 °C oven.
2. Add resin to the previously polymerized BEEM capsules until a slightly convex meniscus is obtained (Fig. 1). Hence, two phases will coexist inside the capsule: the solid one, composed of previously polymerized resin, and the recently added liquid one.
3. Cover the BEEM capsule with a coverslip (Fig. 1), placing the side with the cells in contact with the resin (face down) of the

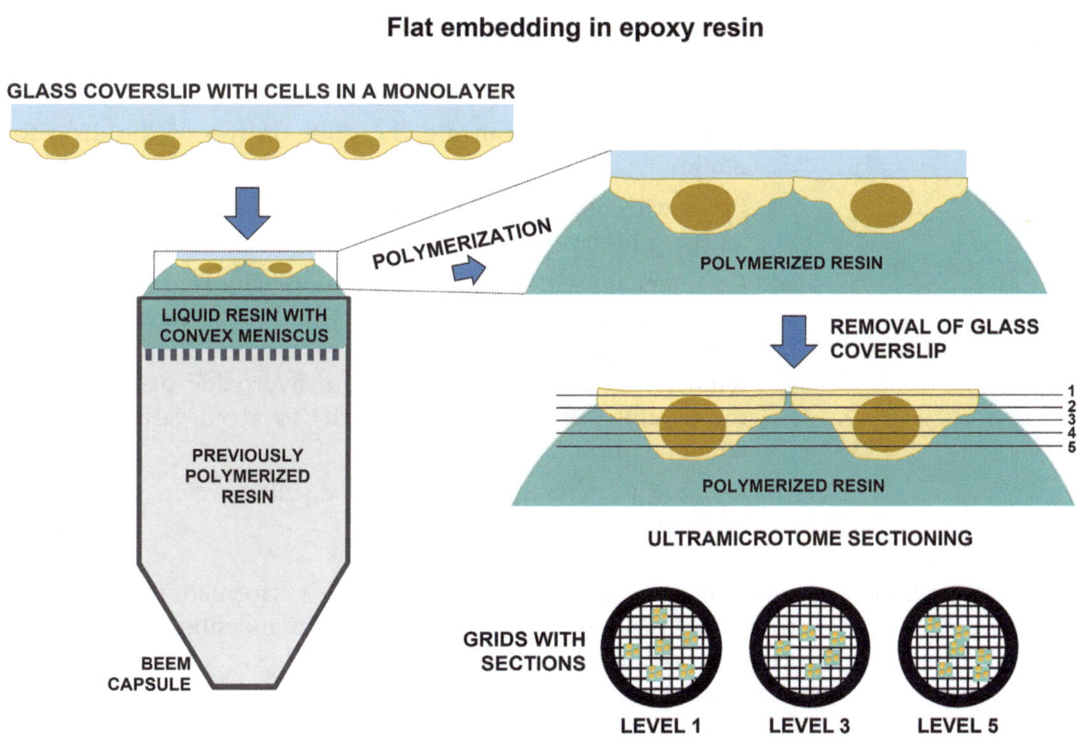

Fig. 1 Flat embedding of cells in resin. The coverslip with cells should be carefully placed on top of the convex meniscus formed by the liquid resin (Subheading 3.1.5, **steps 1–3**). Once the samples have polymerized, the coverslips should be removed to keep the cell layer attached to the polymerized resin block. It is recommended to obtain 4–6 sections to ensure cell detection (Subheading 3.1.6)

BEEM capsule. The coverslip should be centered on the surface of the capsule. Ensure there are no gaps: previously polymerized resin, recently added liquid resin, and coverslip should be in contact without bubbles or voids.

4. Incubate the BEEM capsules in a vertical position, avoiding inclinations, in a 60 °C oven for 3 days for the correct resin polymerization.

3.1.6 Glass Extraction and Sectioning

Once the samples have polymerized, the glass coverslips must be removed so that the layer of cells remains attached to the polymerized resin block (Fig. 1). The glass coverslips are detached using thermal contrast between liquid nitrogen and boiling water (*see* **Note 5**). Fill a polystyrene container with liquid nitrogen and a glass beaker with boiling water. Immerse the BEEM capsule upside down in the liquid nitrogen (ideally only the glass where the cell culture was made) and then in boiling water. Repeat the process until the glass is released by thermal shock. The more volume of the capsule BEEM is submerged, the more likely the polymerized resin is to crack (*see* **Note 6**). For conventional samples from cell culture, ultrathin sections (60–70 nm) should be obtained in an Ultramicrotome. It is advisable to obtain between 4 and 6 sections to assure the detection of cells (Fig. 1) (*see* **Note 7**).

3.1.7 Contrast of Ultrathin Sections

1. Place a drop of UranyLess EM on parafilm.
2. Stain the sections, putting the grids on top of the drop for 2 min.
3. Drain with filter paper, rinse thoroughly in distilled water, and let the grid dry.
4. Counterstain with lead citrate solution: This step is performed in Petri dishes, where a drop of the lead citrate solution is placed on parafilm and the grid on top for 1 min (at room temperature). Place several sodium hydroxide pellets around the parafilm (inside the Petri dish) to avoid carbon dioxide contamination.
5. Drain with filter paper, rinse in distilled water, and let the grid dry.

3.1.8 Grid Storage and Preservation

Once the contrast of ultrathin sections is performed, they are ready to use in TEM. Grids with sections have an indefinite duration and can be stored in respective labeled Petri dishes at room temperature inside a dehumidifier chamber with silica gel.

In various study models, TEM images are commonly used to visualize and confirm lysosomal MCS with other organelles. Accurately identifying structures of interest is critical and challenging, as lysosomal structures from fibroblasts often exhibit a wide range of morphologies in TEM micrographs. The endolysosomal system is a dynamic compartment that constantly forms structures with varying shapes and cargo content based on their composition and the nature of the cell type. This guideline aims to provide simplified but practical directions for identifying structures belonging to the endolysosomal system in human skin fibroblasts. The identification will be done through visual analysis of TEM images, which will be used in the morphometric analysis and MCS quantification.

3.2 Image Acquisition and Analysis: Identification of Lysosomal Structures and Lysosome–Mitochondria Membrane Contact Sites

3.2.1 Initial Considerations for Image Capture

1. Once the grid has been installed into the electron microscope and the cells have been focused (Fig. 2a), search along the cell for lysosomal structures (Fig. 2b–e).
2. The recognition of these structures is based on at least one of the following criteria: (i) high electron density of the structure, (ii) visible delimitating membrane, (iii) rounded shape, (iv) diameter of the structure: in a physiological state, it oscillates between 50 and 500 nm [30, 31]. Once identified, proceed to lysosome size measurement and lysosome–mitochondria MCS localization. For image capture, we estimate that magnification in a range between ×30,000 and ×50,000 is suitable for image analysis. Capture the images as raw data (*see* **Note 8**) and save them in a folder named with cell type, condition, block number, and date. It is important to keep a record of the sections and grids studied to avoid repeated imaging. Consider performing three sets of independent experiments (from cell culture to image analysis) for a complete experimental approach. For each set, it is recommended to capture images of at least 20 cells per condition (*see* **Note 9**).

3.2.2 Lysosomal Size Measurements

Before beginning, create a copy folder for the analysis and quantification of images while keeping the original folder safe to avoid any unwanted modifications to the original files.

1. Open ImageJ software.
2. Click the *Analyze* button and then *Set Measurements.* A new screen will open, and select the *Area* and *Perimeter* boxes while keeping the remaining boxes unselected. Keep the *Redirect* option as *None* and adjust the number of decimals with which you want to work (we recommend three).
3. Open the image of interest in the software by dragging the respective file toward the ImageJ toolbar. The image magnification can be adjusted with the keyboard buttons +/−.
4. Delimit the lysosomal structure (*see* **Note 10**) with the *Polygon Selection* tool until leaving a closed shape (connecting initial and final selection dots). It is convenient to delineate as many

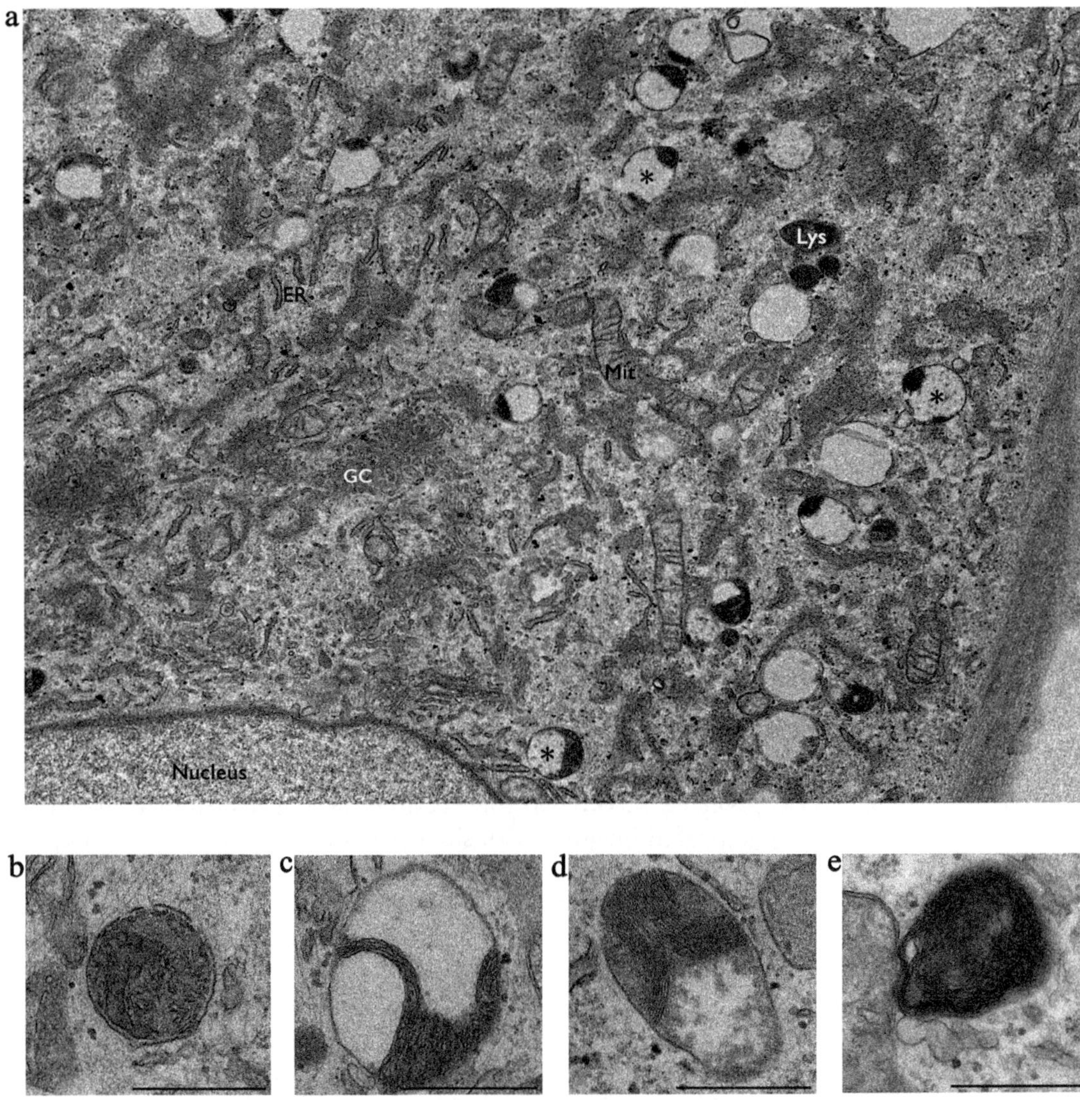

Fig. 2 Low magnification micrograph of flat-embedded HSF showing (**a**) prototypical appearance of cytoplasm including the major compartments. Scale bar: 1 μm. (**b**) Several typologies of frequently observed LE/Lys in HSF. ER: endoplasmic reticulum; GC: golgi complex; Lys: lysosome; Mit: mitochondria. Asterisks indicate endolysosomes (LE/Lys). Scale bar: 0.5 μm

segments as needed to enclose accurately the lysosomal membrane. If a few segments are drawn, there is a risk of generating straight segments that are too long and do not representatively demarcate the curved edge of the membrane.

5. Once the structure (Region of Interest, ROI) is delimited, open ROI Manager and click the *Add* button to incorporate the selected ROI.

6. Selecting the *Show All* and *Labels* boxes makes it possible to enumerate and visualize the entire set of images according to the draw order.
7. Repeat steps three to five until the list of desired structures is completed in the ROI Manager.
8. Select all the ROIs from the ROI Manager list and click the *Measure* button. A new screen will open with Area and Perimeter values from every selected structure. By having a calibrated image, the unit of measurement will correspond with the image (it will appear on the base of the toolbar).
9. We suggest importing and working with the results in a data spreadsheet (e.g., Microsoft Excel) after saving the original file.

3.2.3 MCS Quantification Expressed as a Percentage of Lysosome Perimeter in Contact with Mitochondria

From a geometric point of view, in ultrathin sections, lysosomes exhibit a more homogeneous morphology than mitochondria, which are mainly present as a network within the cell. This is relevant when performing MCS quantification from TEM samples, as the mitochondrial structure in an ultrathin section may not necessarily represent the whole mitochondrial structure, leaving the lysosome as an adequate reference standard to compare the degree of MCS between the two organelles. In view of this, we use the lysosomal perimeter as a reference variable.

1. Once lysosomes are localized, perform MCS identification (Fig. 3a) with mitochondria. To achieve this, follow these criteria: i) the distance between both organelles should be ≤30 nm (*see* **Note 11**). In some cases, one lysosome appears in close contact with more than one mitochondrion, as well as one mitochondrion exhibits MCS with more than one lysosome; those should also be included.
2. After lysosome–mitochondria MCS has been identified, verify the distance between both membranes by drawing a line from limit to limit between membranes using the *Straight-Line* button on the toolbar. The distance value will appear in the Information Bar.
3. Follow **steps 4–8** from Subheading 3.2.2 for lysosomal perimeter measurement (Fig. 3b).
4. Using the *Straight Line* tool, trace the extreme sides (Fig. 3b) where the membrane-to-membrane distance exceeds 30 nm (*see* **Note 11**).
5. Using the *Segmented Line* tool, trace a central segmented line (Fig. 3b) between lines marking the boundaries of the MCS zone traced in **step 4** and add it to the ROI Manager. By doing this, the length of MCS will be obtained.
6. Obtain the lysosomal perimeter and length of MCS by clicking the *Measurement* button.

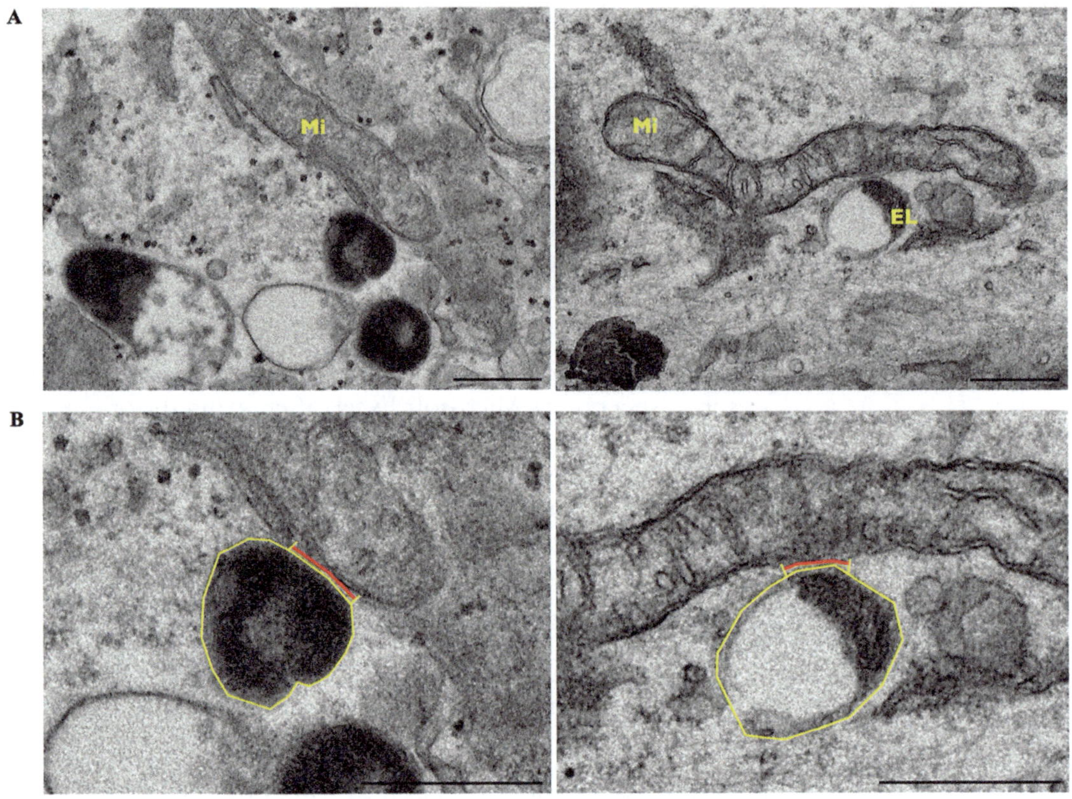

Fig. 3 LE/Lys-Mitochondria Membrane Contact Sites (MCS) analysis in HSF (**a**) Examples of MCS between a lysosome (Lys) (*left*) and endolysosome (LE/Lys) (*right*) with mitochondria. Scale Bar: 0.5 μm. (**b**) Analysis of MCS extent in images magnified from (**a**). The colored tracing corresponds to section 3.2.3, which shows the delimitation of the lysosomal structure in yellow (**step 3**), the enclosure of the MCS region in yellow (**step 4**) and, the line that spans the MCS length in red (**step 5**). LE/Lys: endolysosome, Lys: Lysosome, Mit: Mitochondria. Scale bar: 0.5 μm

7. We recommend importing and working with the results in a data spreadsheet (e.g., Microsoft Excel) after saving the original file.
8. Calculate the percentage of endolysosome/lysosome in contact with mitochondria using the formula (*see* **Note 12**):

$$\%\text{contact} = \frac{\text{Lenght of MCS} \times 100}{\text{Lysosome Perimeter}}$$

9. Results should be plotted as the percentage of the lysosomal perimeter in contact with mitochondria, allowing for a comparison between different experimental conditions.

4 Notes

1. Osmium tetroxide must be prepared under a hood and with gloves due to its hazardous nature. It is important to note that osmium tetroxide vapors can penetrate plastics; therefore, it should be stored in a sealed glass container with unbreakable secondary containment.
2. The minimum fixation time is 1 h at room temperature or 2 h at 4 °C. In the latter case, changing the solution after half an hour of fixation is advisable.
3. With appropriate coverslip storage, cells can last for several weeks. To recover and continue to post-fixation steps, perform a couple of washes with the fixative solution.
4. The epoxy resin is a hazardous substance. Hence, handle it in the fume hood with gloves and a mask. Different consistencies of the solution can be achieved depending on the ratio of the different components. The proportions shown in this protocol are to obtain the most commonly used one. Avoid the formation of bubbles.
5. Insulating gloves and safety glasses must be used when handling liquid nitrogen. When a resin block with a coverslip is exposed to highly opposite temperatures, bubbles and jumps are common, so it must always be handled with care to avoid splashes.
6. The surface of the BEEM capsule where the glass was attached must be completely smooth and shiny, like a mirror. As an optional, using a magnifying glass or eyepieces of a pyramidotome, play with the angle of incidence of the light to see the black texture that corresponds to the sample.
7. Fibroblasts are approximately 3–4 μm in height once fixed and processed (embedding), which theoretically allows for approximately 40 sections (70 nm) to be obtained. However, since 4–6 sections of 60–70 nm per grid are cut, the analysis of 3–5 levels is sufficient to obtain a representative view of the volumetric distribution of organelles within the cell.
8. It is relevant to store the images captured in the format provided by the software to facilitate measurement calibration during image analysis. By having a calibrated image, the units of measurement will correspond with the image being processed in ImageJ (it will appear at the base of the toolbar).
9. HSFs have a large cell surface area (±6000 μm^2). Hence, using at least x30,000 magnification for image capture will involve capturing several images to cover the entire cell surface. Therefore, we recommend that for each cell, priority should be given

to capturing images of those areas with the lysosomal structures of interest.

10. In some cases, the membrane of the lysosome (or mitochondria) is not well-defined due to imperfections inherent to sample handling (or the angle of sectioning). In those cases, and if the structure is correctly identified, perform the measurement considering the blurred border as the limit of the structure.
11. Although the consensus definition of MCSs specifies a maximum distance of 30 nm (and a minimum of 10 nm) between two organelle membranes, recent literature suggests extending this value to as much as 80 nm [32] due to the dynamic nature of membrane interactions. It should also be pointed out that the TEM images offer a two-dimensional picture of a three-dimensional process. Hence, we recommend not ruling out a possible MCS if the distance between the two membranes falls within the range of 30 nm to 50 nm.
12. Be aware that the Length of MCS and Perimeter of Lysosomes variables should be expressed with the same unit of measurement (commonly micrometers) to obtain the correct percentage result.

Acknowledgments

This work was supported by Instructor Scholarship from the Graduate School and the Biological Sciences Faculty, both from Pontificia Universidad Católica de Chile (CPR); the Ministerio de Ciencia e Innovación (PID2020-115910RB-I00), España (CE); and Fondecyt 1230337 (Agencia Nacional de Investigación y Desarrollo, Chile) (SZ).

We are indebted to Eva Prats and all the technical staff from Centres Científics i Tècnics/University of Barcelona (CCiT/UB), Electron Microscopy Unit (Casanova/Clínic), for the help and skillful assistance. Also, we are grateful to Prof. Emily Eden (University College, London, United Kingdom) for helpful advice with the flat embedding procedure and to Marc Bernaus-Esqué for his assistance in the preparation of cell culture.

References

1. Porter KR, Palade GE (1957) Studies on the endoplasmic reticulum. III. Its form and distribution in striated muscle cells. J Biophys Biochem Cytol 3(2):269–300. https://doi.org/10.1083/jcb.3.2.269
2. Vance JE (1990) Phospholipid synthesis in a membrane fraction associated with mitochondria. J Biol Chem 265(13):7248–7256
3. Cisneros J et al (2022) Mitochondria-lysosome contact site dynamics and misregulation in neurodegenerative diseases. Trends Neurosci

45(4):312–322. https://doi.org/10.1016/j.tins.2022.01.005

4. Cremer T, Neefjes J, Berlin I (2020) The journey of Ca(2+) through the cell – pulsing through the network of ER membrane contact sites. J Cell Sci 133(24). https://doi.org/10.1242/jcs.249136
5. Phillips MJ, Voeltz GK (2016) Structure and function of ER membrane contact sites with other organelles. Nat Rev Mol Cell Biol 17(2):69–82. https://doi.org/10.1038/nrm.2015.8
6. Vance JE (2014) MAM (mitochondria-associated membranes) in mammalian cells: lipids and beyond. Biochim Biophys Acta 1841(4):595–609. https://doi.org/10.1016/j.bbalip.2013.11.014
7. Voelker DR (2009) Genetic and biochemical analysis of non-vesicular lipid traffic. Annu Rev Biochem 78:827–856. https://doi.org/10.1146/annurev.biochem.78.081307.112144
8. Prinz WA, Toulmay A, Balla T (2020) The functional universe of membrane contact sites. Nat Rev Mol Cell Biol 21(1):7–24. https://doi.org/10.1038/s41580-019-0180-9
9. Wang Z, Benning C (2012) Chloroplast lipid synthesis and lipid trafficking through ER-plastid membrane contact sites. Biochem Soc Trans 40(2):457–463. https://doi.org/10.1042/BST20110752
10. Wenzel EM, Elfmark LA, Stenmark H, Raiborg C (2022) ER as master regulator of membrane trafficking and organelle function. J Cell Biol 221(10). https://doi.org/10.1083/jcb.202205135
11. Wong YC, Kim S, Peng W, Krainc D (2019) Regulation and function of mitochondria-lysosome membrane contact sites in cellular homeostasis. Trends Cell Biol 29(6): 500–513. https://doi.org/10.1016/j.tcb.2019.02.004
12. Ballabio A, Bonifacino JS (2020) Lysosomes as dynamic regulators of cell and organismal homeostasis. Nat Rev Mol Cell Biol 21(2): 101–118. https://doi.org/10.1038/s41580-019-0185-4
13. Belton TB, Leisten ED, Cisneros J, Wong YC (2022) Live cell microscopy of mitochondria-lysosome contact site formation and tethering dynamics. STAR Protoc 3(2):101262. https://doi.org/10.1016/j.xpro.2022.101262
14. Leisten ED, Woods AC, Wong YC (2023) Super-resolution microscopy: insights into mitochondria-lysosome crosstalk in health and disease. J Cell Biol 222(12). https://doi.org/10.1083/jcb.202305032
15. Wong YC, Ysselstein D, Krainc D (2018) Mitochondria-lysosome contacts regulate mitochondrial fission via RAB7 GTP hydrolysis. Nature 554(7692):382–386. https://doi.org/10.1038/nature25486
16. Pickles S, Vigie P, Youle RJ (2018) Mitophagy and quality control mechanisms in mitochondrial maintenance. Curr Biol 28(4):R170–R185. https://doi.org/10.1016/j.cub.2018.01.004
17. Sugiura A, McLelland GL, Fon EA, McBride HM (2014) A new pathway for mitochondrial quality control: mitochondrial-derived vesicles. EMBO J 33(19):2142–2156. https://doi.org/10.15252/embj.201488104
18. Horvath SE, Daum G (2013) Lipids of mitochondria. Prog Lipid Res 52(4):590–614. https://doi.org/10.1016/j.plipres.2013.07.002
19. Arruda AP et al (2014) Chronic enrichment of hepatic endoplasmic reticulum-mitochondria contact leads to mitochondrial dysfunction in obesity. Nat Med 20(12):1427–1435. https://doi.org/10.1038/nm.3735
20. Enrich C et al (2021) Annexins bridging the gap: novel roles in membrane contact site formation. Front Cell Dev Biol 9:797949. https://doi.org/10.3389/fcell.2021.797949
21. Enrich C et al (2019) Cholesterol overload: contact sites to the rescue! Contact (Thousand Oaks) 2:2515256419893507. https://doi.org/10.1177/2515256419893507
22. Solsona-Vilarrasa E et al (2019) Cholesterol enrichment in liver mitochondria impairs oxidative phosphorylation and disrupts the assembly of respiratory supercomplexes. Redox Biol 24:101214. https://doi.org/10.1016/j.redox.2019.101214
23. Torres S et al (2017) Lysosomal and mitochondrial liaisons in Niemann-pick disease. Front Physiol 8:982. https://doi.org/10.3389/fphys.2017.00982
24. Hoglinger D et al (2019) NPC1 regulates ER contacts with endocytic organelles to mediate cholesterol egress. Nat Commun 10(1):4276. https://doi.org/10.1038/s41467-019-12152-2
25. Meneses-Salas E et al (2020) Annexin A6 modulates TBC1D15/Rab7/StARD3 axis to control endosomal cholesterol export in NPC1 cells. Cell Mol Life Sci 77(14):2839–2857. https://doi.org/10.1007/s00018-019-03330-y
26. Kim S, Wong YC, Gao F, Krainc D (2021) Dysregulation of mitochondria-lysosome contacts by GBA1 dysfunction in dopaminergic neuronal models of Parkinson's disease. Nat

Commun 12(1):1807. https://doi.org/10.1038/s41467-021-22113-3

27. Onukwufor JO, Dirksen RT, Wojtovich AP (2022) Iron dysregulation in mitochondrial dysfunction and Alzheimer's disease. Antioxidants (Basel) 11(4). https://doi.org/10.3390/antiox11040692
28. Peng W, Wong YC, Krainc D (2020) Mitochondria-lysosome contacts regulate mitochondrial Ca(2+) dynamics via lysosomal TRPML1. Proc Natl Acad Sci USA 117(32): 19266–19275. https://doi.org/10.1073/pnas.2003236117
29. Schindelin J et al (2012) Fiji: an open-source platform for biological-image analysis. Nat Methods 9(7):676–682. https://doi.org/10.1038/nmeth.2019
30. Bandyopadhyay D et al (2014) Lysosome transport as a function of lysosome diameter. PLoS One 9(1):e86847. https://doi.org/10.1371/journal.pone.0086847
31. de Araujo MEG, Liebscher G, Hess MW, Huber LA (2020) Lysosomal size matters. Traffic 21(1):60–75. https://doi.org/10.1111/tra.12714
32. Scorrano L et al (2019) (2019) coming together to define membrane contact sites. Nat Commun 10:1287. https://doi.org/10.1038/s41467-019-09253-3

Chapter 11

Assessment of Lysosome Trafficking in iPSC-Derived Neurons Utilizing Time-Lapse Videos and TrackMate ImageJ Plugin Software

Oluwatobi N. Eboda, Indranil Basak, and Stephanie M. Hughes

Abstract

In order to maintain cell homeostasis, lysosomes must be trafficked throughout the cell and associated with different compartments for the delivery and breakdown of macromolecules. This process is especially important in polar cells, such as neurons, that require the trafficking of lysosomes to distal locations to perform their functions. Here, we employed LysoTracker DND-99, a live-cell fluorometric dye that stains acidic organelles, to label lysosomes prior to time-lapse imaging on a Nikon A1 confocal microscope and subsequent assessment of total distance travelled using the TrackMate ImageJ plugin.

Key words Lysosome trafficking, TrackMate, ImageJ, LysoTracker

1 Introduction

Lysosomes are acidic, membrane-bound organelles carrying numerous hydrolytic enzymes. They preside at the end of the autophagosome-lysosomal pathway and endocytic pathway and take part in maintaining cellular homeostasis by mediating such processes as lipid homeostasis, energy metabolism, plasma membrane repair, and pathogen defense [1]. Lysosomes maintain an acidic environment of pH 4.5–5.0, which facilitates the activity of the encapsulated hydrolases for the breakdown of macromolecules delivered via endocytosis, phagocytosis, or fusion with autophagic compartments. The trafficking of lysosomes to different areas of the cell is imperative to facilitate proper degradation of waste material, localized recycling, and resource management and to carry out other functions like cargo delivery. This function is especially important in neurons, a cell type with long processes that are

Authors Oluwatobi Eboda and Indranil Basak are co-first author

Stephanie M. Hughes and Indranil Basak (eds.), *Lysosomes: Methods and Protocols*, Methods in Molecular Biology, vol. 2976, https://doi.org/10.1007/978-1-0716-4844-5_11,

quite distal to the cell soma. The neurons require delivery of functional lysosomes down long axonal and dendritic processes to allow for waste management and the recycling of synaptic proteins and vesicles, ensuring proper synaptic function at these distal locations [2–6]. To date, there have been several examples of lysosomal storage disorders showing the pathologic modulation of the trafficking of lysosomes, including in cases with variations in CLN3, CLN5, and CLN8 genes leading to Batten disease and variations in NPC1 and NPC2 genes causing Nieman–Pick disease type C [5, 7–9]. The dysfunction in lysosomal trafficking observed in these conditions highlights the need to assess this phenotype in the research of lysosomal storage disorders, especially those implicated in neurodegeneration. The tracking of lysosomes can be done using several methods; however, the protocol presented here offers the advantage of automated detection and tracking of signals after entering detection parameters. The true power of this methodology is that it facilitates the generation of a large volume of data points versus the manual selection required in other methodologies. This protocol provides an overview of the use of LysoTracker and the TrackMate ImageJ plugin to assess lysosomal trafficking features in neurons.

2 Materials

1. BrainPhys cortical neuron media: Cortical neuron media is prepared in 50-mL aliquots under aseptic conditions for use within 7–10 days. 49 mL of BrainPhys media (+/− phenol red), 1 mL of B27 (50X) supplement, 50 μL of BDNF (10 μg/mL), 50 μL of Laminin (1 mg/mL), and 50 μL of NT3 (10 μg/mL). The media is warmed to 37 °C before use.
2. 8-well chambered coverslips (Nunc Lab-Tek II Chamber Coverglass).
3. LysoTracker™ Red DND-99 dye.
4. Hoechst 33342 nuclear stain.
5. G3 mNGN2 dCas9 Day 0 i^{3}Ns [10, 11].
6. Nikon A1+ Inverted Confocal Laser Scanning Microscope.
7. ImageJ [12] and Trackmate [13, 14].
8. GraphPad Prism.

3 Methods

3.1 Plating

1. G3 mNGN2 dCas9 Day 0 i^{3}Ns are generated from induced pluripotent stem cells carrying an integrated mouse neurogenin-2 transcription factor transgene under the control

of doxycycline doxycycline-inducible promoter. This allows for the generation of a pure population of isogenic, integrated, inducible neurons (i^{3}Ns). Plate Day 0 i^{3}Ns at a density of 1×10^5 cells per well to 8-well chambered coverslips in 500 μL of cortical neuron media.

2. Maintain i^{3}Ns for 14 days, half-media changes every 3 days with pre-warmed cortical neuron media. Detailed protocol describing the generation and culture of these cells can be found in the Basak et al. and Fernandopulle et al. [7, 10].

3.2 Lysotracker

1. Dilute LysoTracker 1 mM stock solution to 1:20,000, via 1: 100 then 1:200 step dilutions, in cortical neuron culture media to a final concentration of 50 nM. Protect from light. Make up to the final volume necessary for the entire experiment.
2. Transport culture plate to cell culture hood, remove culture media, and replace with up to 500 μL of LysoTracker media mix, or appropriate well volume (24-well: 0.5 mL, 6-well: 1–2 mL, 10-cm dish: 10 mL), then return plate to a 37 °C cell culture incubator for 30 min.
3. During the incubation period, make up 1 μg/mL of Hoechst 33342 nuclear stain in phenol red-free cortical neuron media. Protect from light.
4. After incubation, return the plate to the cell culture hood and remove LysoTracker media, and replace with phenol red-free cortical neuron media with Hoechst 33342 nuclear stain in phenol red-free cortical neuron media from **step 3** and allow to incubate for 10 min in the cell culture hood at room temperature. Protect from light.
5. After 10 min, remove Hoechst media and wash twice gently with PBS before replacing it with warm phenol red-free cortical neuron media. Image immediately.

3.3 Imaging

1. Generate time-lapse videos using a 60X objective with 512-pixel image quality to minimize image intervals for tracking analysis. Use any confocal microscope capable of simultaneous time-lapse imaging in multiple channels (405 nm, 488 nm, 555 nm, 647 nm). Videos were generated in two channels, 405 nm and 555 nm. i^{3}Ns express a blue fluorescent protein (BFP) cassette and fluoresce in the 405 nm channel for visualization of neural morphology. This property was taken advantage of when imaging, while LysoTracker™ was imaged in the 555-nm channel.
2. Generate a minimum of 3–5× time-lapse videos per well for each chambered coverslip.
3. Save video files in ND2 format, or other formats with embedded metadata.

3.4 TrackMate Plugin

1. Open the ImageJ program.
2. Open the first video file, and the options window of the FIJI embedded plugin Bio-formats Importer will pop up. If this does not occur, manually open the importer to choose how metadata is loaded into imaging software; click split channels.
3. In ImageJ: Image > Color > Merge Channels > Assign channels and create composite image > Save Image.
4. Select the composite image, and open TrackMate. If not already part of the ImageJ suite, the plugin can be downloaded and installed from here: https://imagej.net/plugins/trackmate/

   Plugins > Tracking > TrackMate
5. TrackMate pop-up window will open, displaying image metadata, as seen in Fig. 1.
6. Click "Next" (in Fig. 1); the next window will prompt the user to select a detector from the drop-down menu. Figure 2 displays the TrackMate window. Detector should be selected based on the user's need in accordance with their images.

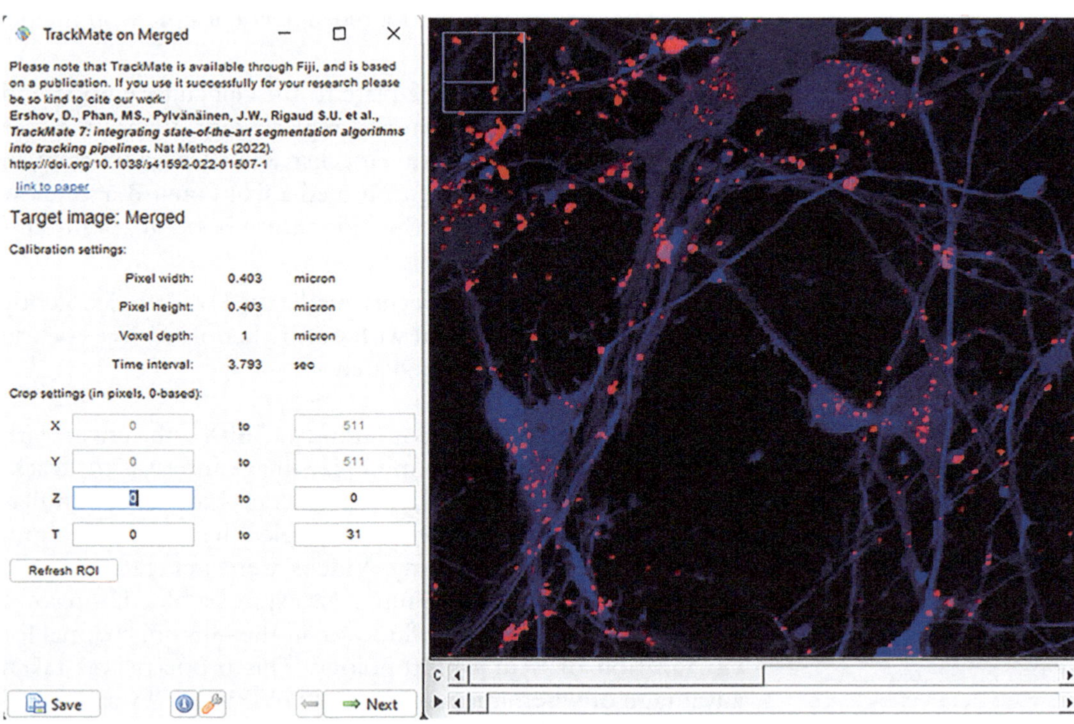

Fig. 1 When opening the TrackMate Image J plugin on a sequential image file, the plugin reads embedded metadata to fill in pixel width and height, voxel depth, time interval, and other data presented here. Saving the original files as an .nd2 format allows for the plugin to use this metadata, but any file format with embedded metadata should produce the same results

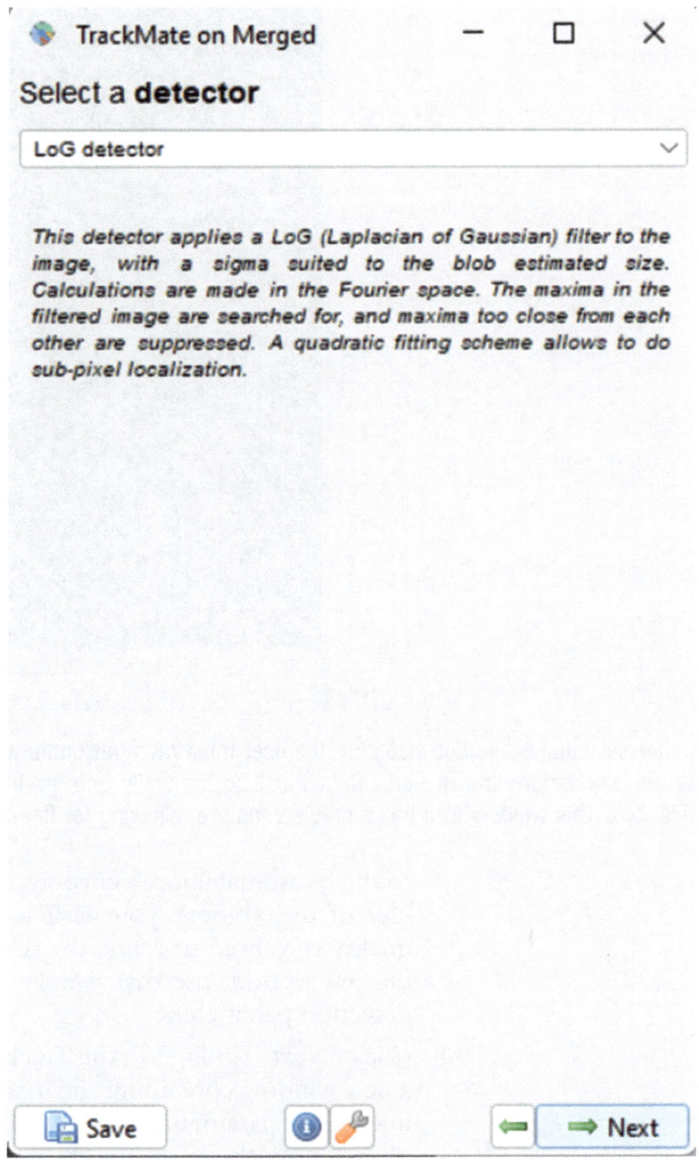

Fig. 2 Plugin provides a brief description of the selected detector. End users should evaluate different selectors and use what is most appropriate for their dataset. In this protocol, the Laplacian of Gaussian detector was used to identify lysosomes

Trackmate plugin provides a brief description of each selector. In this application, the LoG detector is used to detect "blobs" of the estimated size with sub-pixel localization.

7. Click "Next" (in Fig. 2); in the LoG detector window (Fig. 3), set "Detect in channel," "Estimated object diameter," and "Quality Threshold." Channel options will be available for each channel in the image; select the channel corresponding

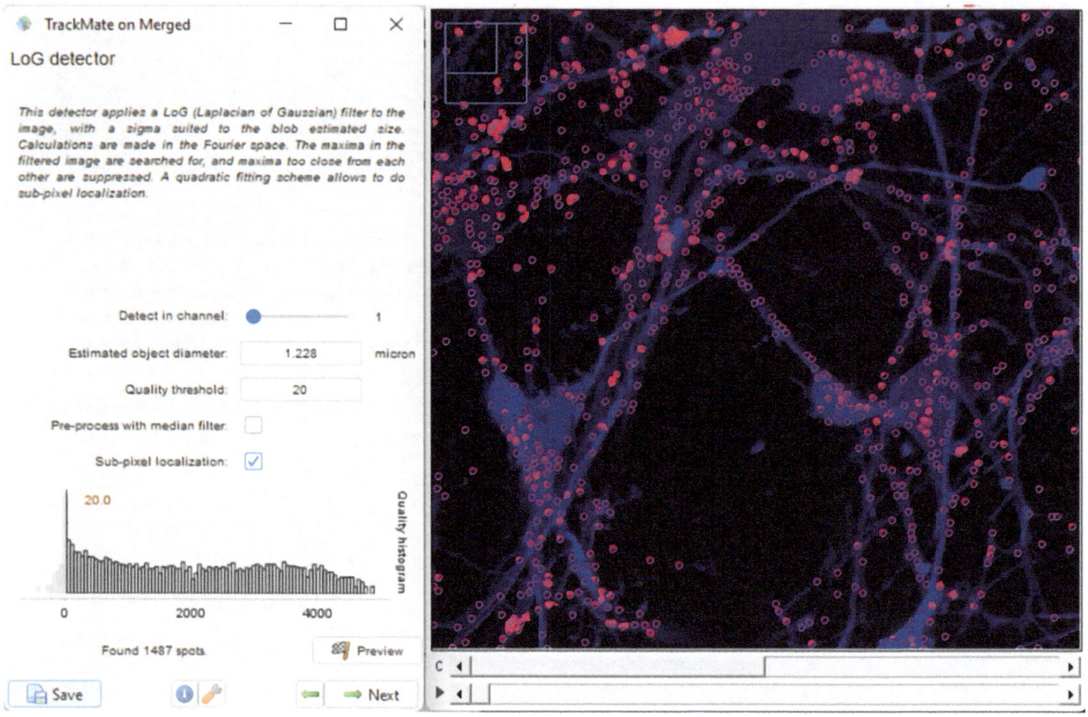

Fig. 3 After selection of the LoG detector, the user must then designate which channel to apply the filter as well as the parameters shown here. Data must be manually entered into the estimated object diameter numerical box. This window also has a preview feature, allowing for fine-tuning of selection criteria

to the lysosomal fluorescence signal. Enter the estimated diameter of the "blobs" you wish to detect and set a minimum quality threshold to refine the selection. This window offers a preview option; use that feature to your advantage in setting detection parameters.

8. Click "Next" (in Fig. 3); the TrackMate plugin will then display a new window containing the macro to detect all spots according to set parameters. This window will only appear as the plugin runs the macro code in the background. Click next when complete. Subsequent steps will allow the user to apply additional filters to refine the selection (*see* **Note 1**).
9. Click "Next"; perform initial thresholding for quality checks of selected spots before other features are rendered. Adjust as needed (Fig. 4).
10. Click "Next" (in Fig. 4); add/remove filters on spots options given in Fig. 5. A range of different filters can be set from the drop-down menu and adjusted accordingly to refine their selection.
11. Click "Next" (in Fig. 5); in the next window (Fig. 6), select a tracker from the drop-down menu; the TrackMate plugin will

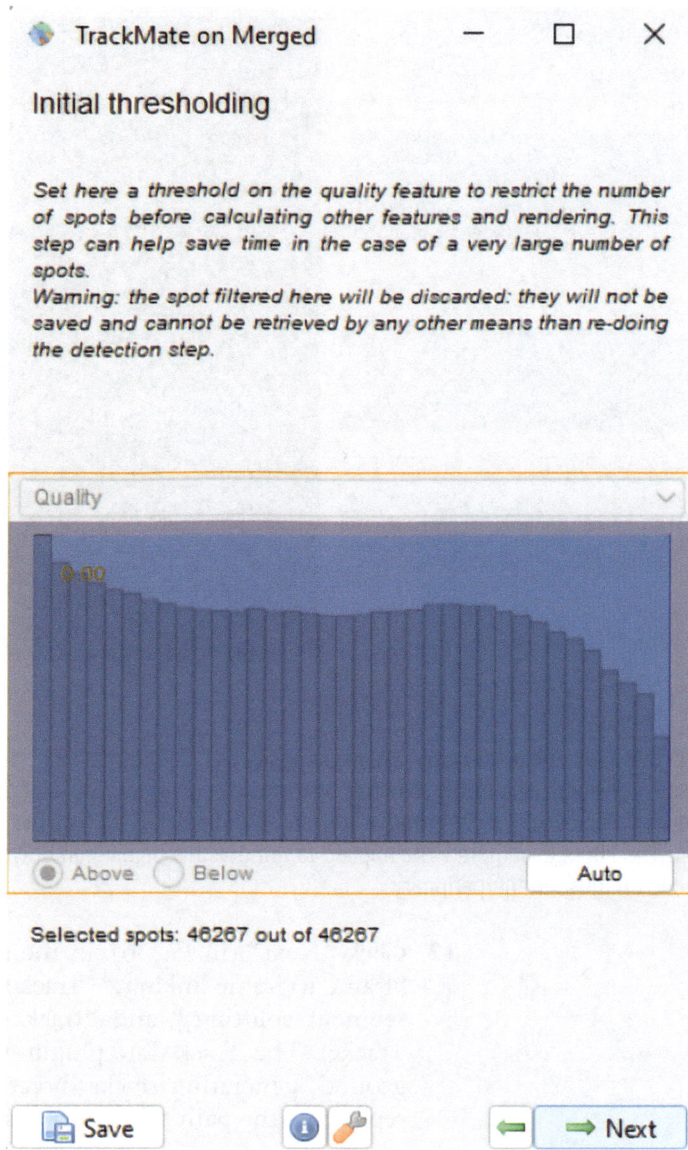

Fig. 4 Initial thresholding on spot detection quality. Spots removed at this step will have no further features calculated, and spot data will be lost

provide detailed descriptions of how each tracker detects spots and builds subsequent tracks; choose one that best suits the data you are working with. In this application, the Linear Assignment Problem (LAP) Tracker was used for 2D planar tracking of particles. The LAP Tracker operates in two steps: first, it performs frame-by-frame linking to build track segments; second, it examines these segments for gap-closing, merging, and splitting events.

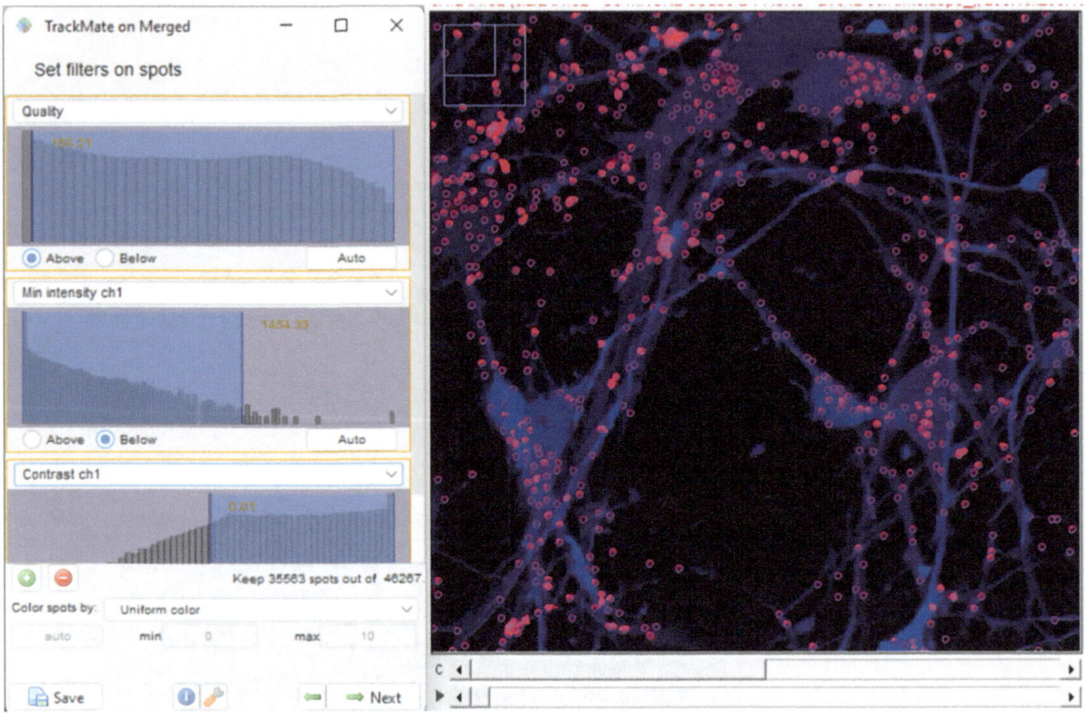

Fig. 5 In this application, "Quality," "Mean Intensity ch1," "Contrast ch1" were used to select lysosomes in the periphery. Cell soma localized lysosomes were more aggregated, lacked defined borders, were brighter due to a collective fluorescent signal, and were largely immobile, making the selection of individual lysosomes difficult. The filters used here were applied to remove this population of lysosomes that could not be isolated and focus on those in the periphery

12. Click "Next" (in Fig. 6); in the next window (Fig. 7), adjust "frame to frame linking," "track segment gap closing," "track segment splitting," and "track segment merging" for LAP Tracker. The TrackMate plugin will run a macro in the background, generating tracks directly on the image. Each track represents the path of an individual lysosome.
13. Click "Next" (Fig. 7); the following window (Fig. 8) presents options to add/remove filters to fine-tune the track population. "Number of spots in track" filter was used in this application to remove spots that were detected in only a single instance in a frame.
14. Click "Next" (Fig. 8); the Display options window (Fig. 9a, top) will appear. Ensure that the checkboxes for "Display spots" and "Display tracks" are selected. Click "Track Scheme" (Fig. 9a) to open a separate window displaying a frame-by-frame "scheme" of all detected spots in the image (Fig. 9b).
15. In the same window (Fig. 9a), the "Tracks" function can be clicked to open a separate window (Fig. 10) containing all collected track feature data.

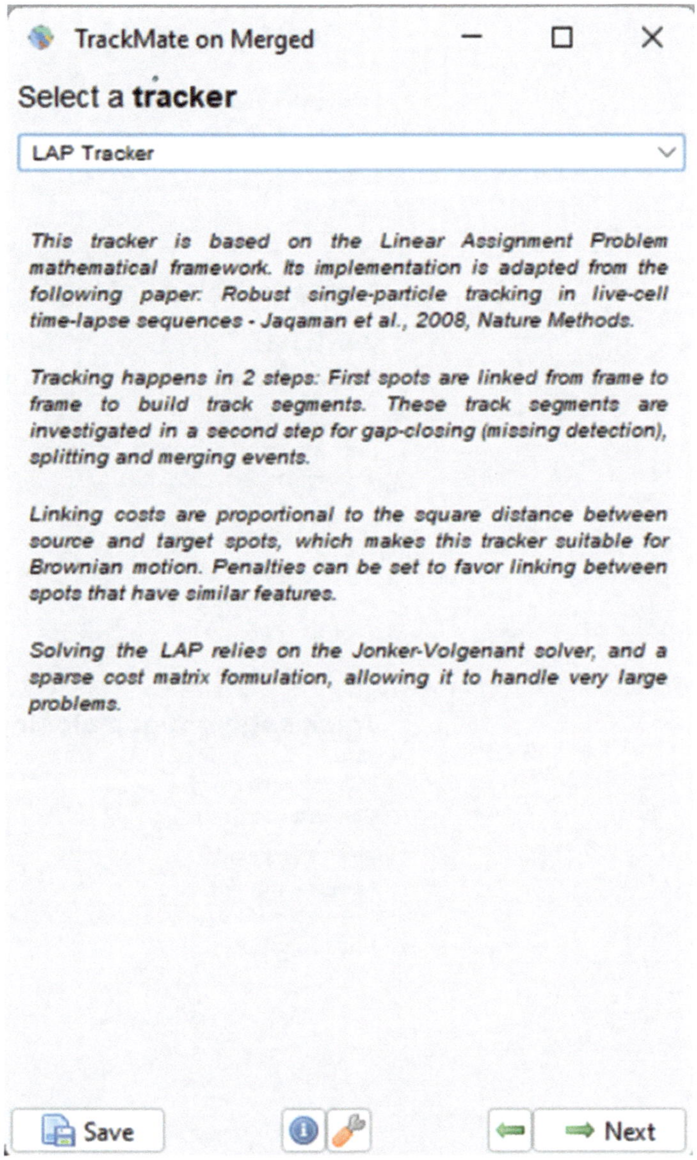

Fig. 6 The LAP Tracker was used in this protocol. The LAP Tracker is based on the Linear Assignment Problem mathematical framework, and its application is described in Jaqaman et al., "Robust single-particle tracking in live-cell time-lapse sequences," Nature Methods, 2008

16. Click "Next" x2 (Fig. 9a); the final window (Fig. 11) will prompt users to select an action from the drop-down window. The Trackmate plugin provides a brief description of what each action does. In this application, "Capture Overlay" was used to generate a video file with embedded spot and track overlays (Fig. 11).

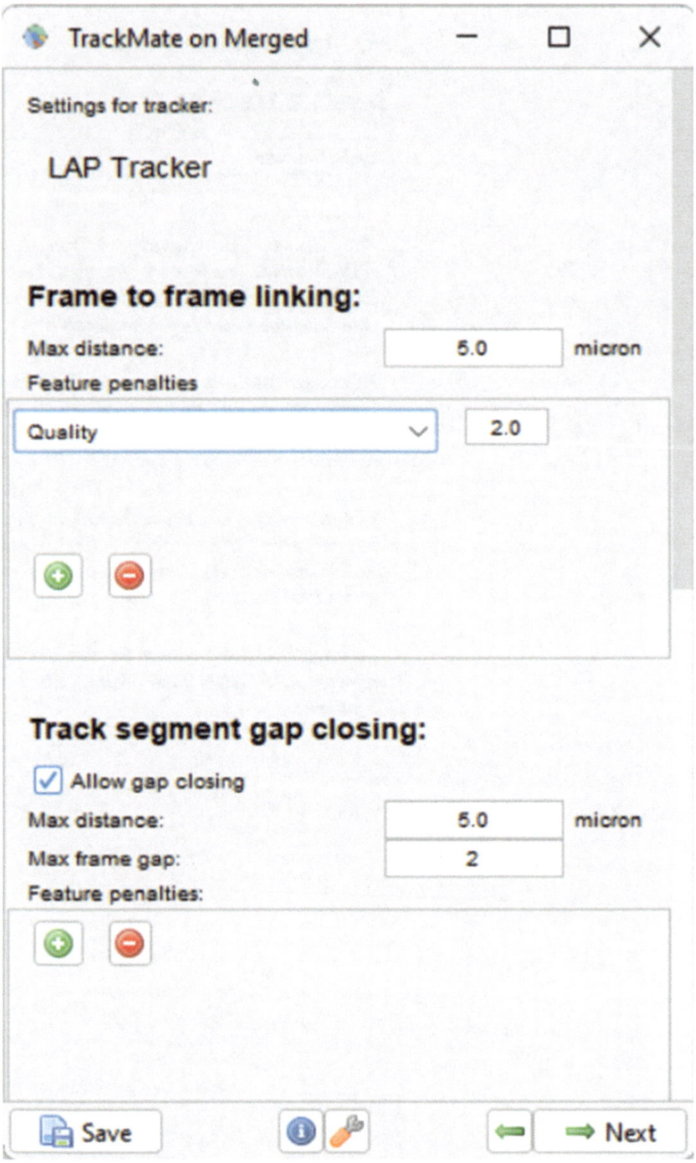

Fig. 7 For this application, "Frame to frame linking" and "Track segment gap closing" were adjusted to produce lysosomal tracks that were manually verified as a final quality check at the end of the protocol

17. Close the window and repeat on all files. The TrackMate plugin automatically creates a text file containing the last used settings that will be loaded when the next image is opened.

3.5 Consolidate and Analyze the Data

1. The TrackMate plugin generates separate .csv files for each time-lapse video processed. Open each and extract relevant data into an Excel file, organized to needs of the user. In this

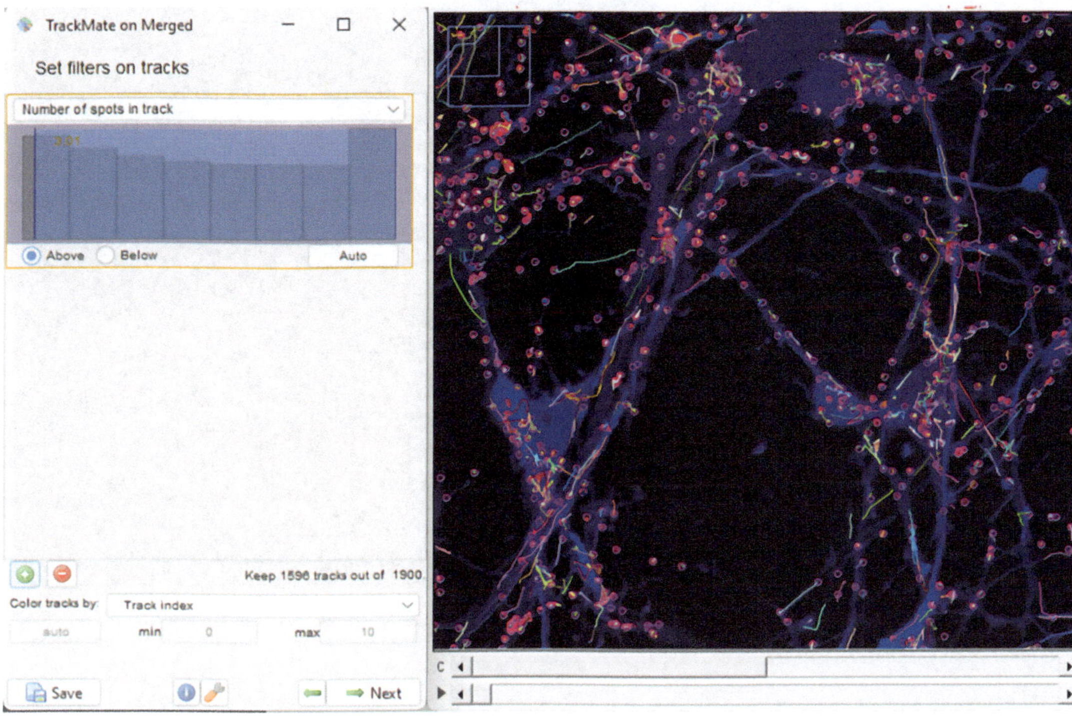

Fig. 8 In this protocol, the "Number of spots in track" filter was used to remove tracks with less than ~3 spots detected. This is not indicative of nonmotile lysosomes but instead lysosomes that do not remain in the field of view long enough to generate a track, or a "blip." Important to note, neuronal processes do not remain on a single plane and extend both below and above the field of view

application, total distance travelled metrics for each "track" were extracted and consolidated into separate Excel files for further processing (*see* **Note 2**).

2. Data analysis was carried out with GraphPad Prism (*see* **Notes 3** and **4**).

4 Notes

1. Initial thresholding of spot quality before spot feature calculation is not recommended in this protocol. Thresholding here removes spot detection data, which is not recoverable. There will be more opportunities for quality thresholding further into the protocol, so the user may choose not to threshold at this point to maintain high data acquisition.
2. The TrackMate plugin provides the user with an abundance of information on each spot/track. For this application, we were interested in assessing whether inhibition of our gene of interest affected the ability of lysosomes to be trafficked to distal locations such as the axons. For this reason, total distance

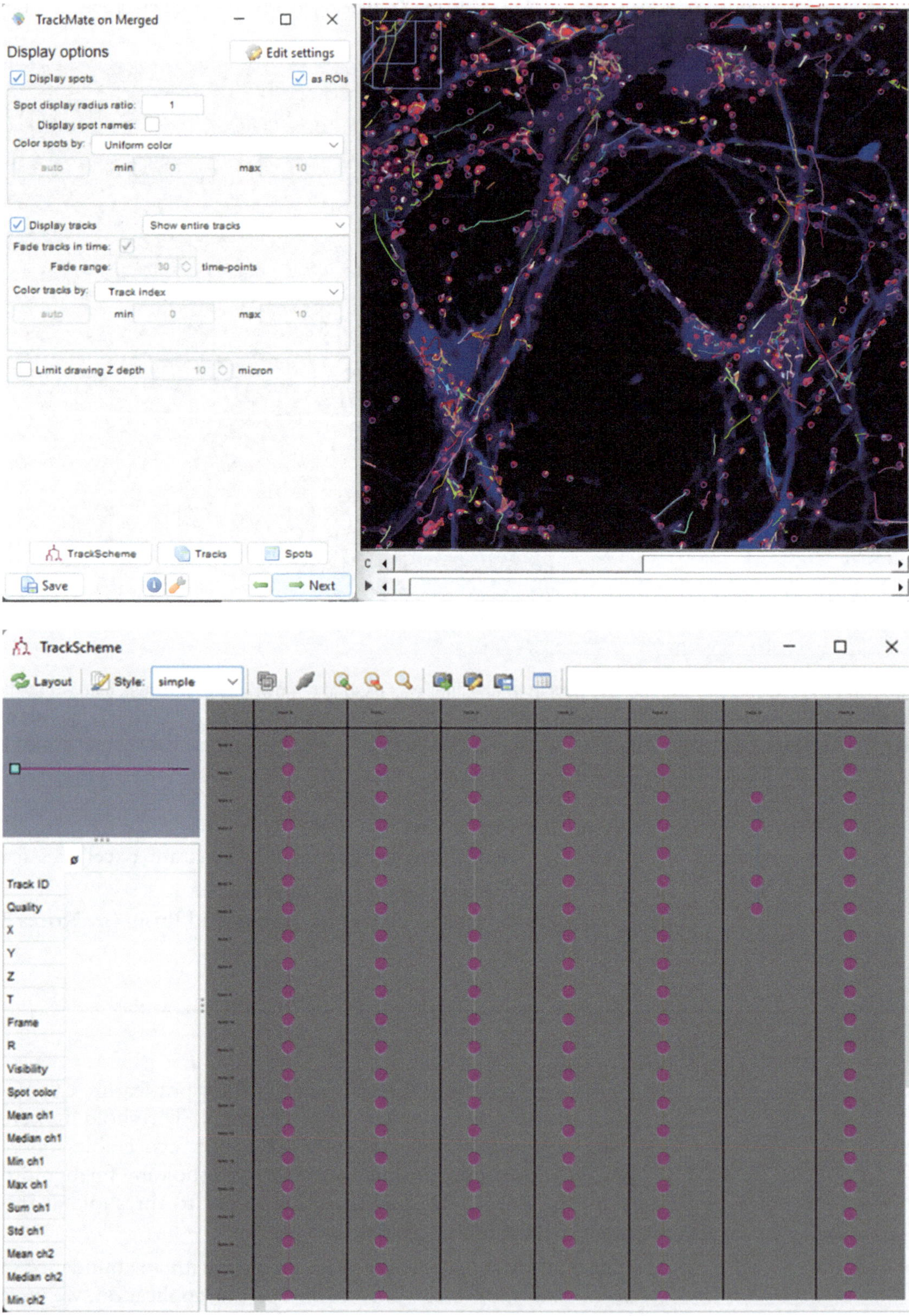

Fig. 9 In the track scheme window, selecting a spot/track will highlight the corresponding spot/track on the image (**a**). Manual track adjustments can be made in this window (**b**), i.e., adding spots where the software lost track, connecting gaps, branching paths, etc. Manual correction and validation of tracks is recommended at this step

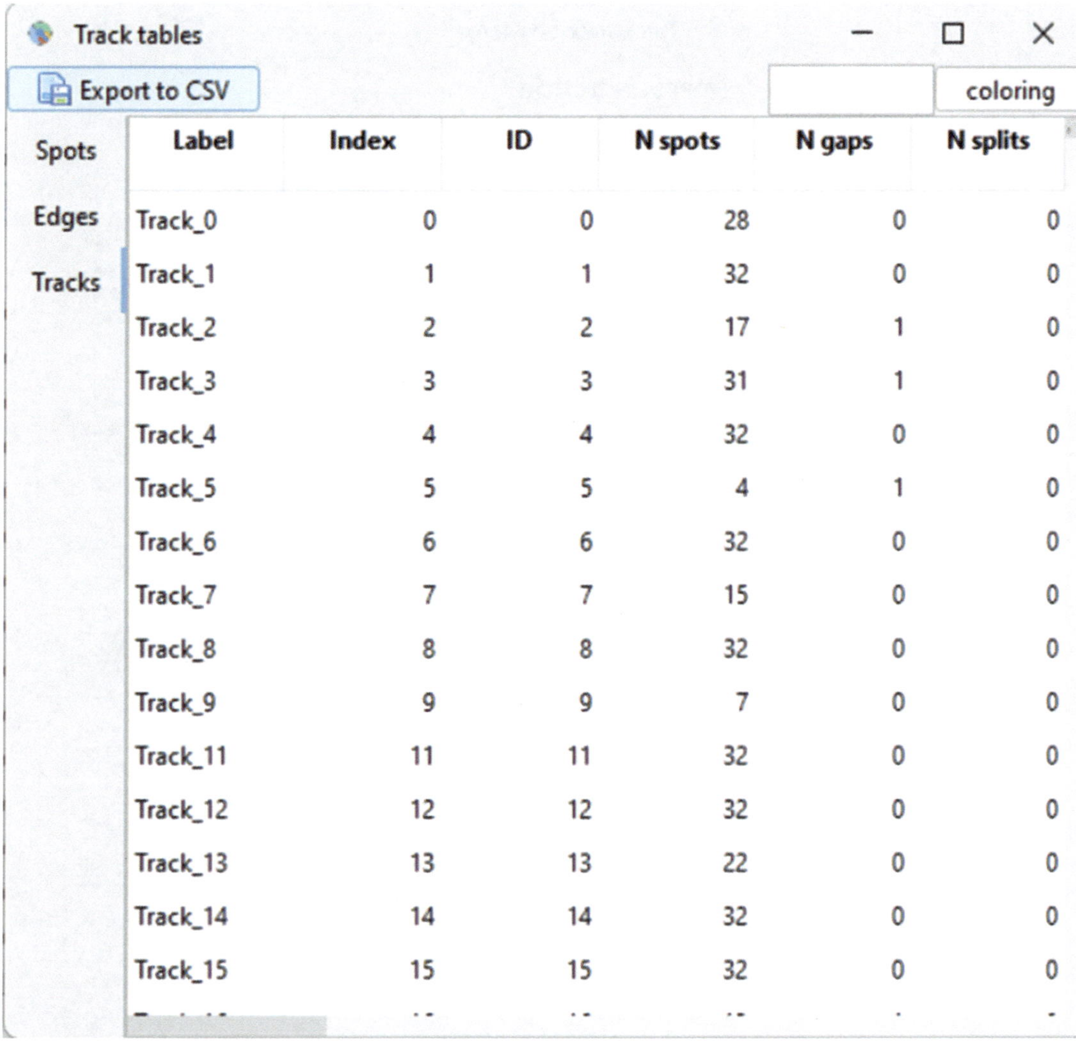

Label	Index	ID	N spots	N gaps	N splits
Track_0	0	0	28	0	0
Track_1	1	1	32	0	0
Track_2	2	2	17	1	0
Track_3	3	3	31	1	0
Track_4	4	4	32	0	0
Track_5	5	5	4	1	0
Track_6	6	6	32	0	0
Track_7	7	7	15	0	0
Track_8	8	8	32	0	0
Track_9	9	9	7	0	0
Track_11	11	11	32	0	0
Track_12	12	12	32	0	0
Track_13	13	13	22	0	0
Track_14	14	14	32	0	0
Track_15	15	15	32	0	0

Fig. 10 Export data table to CSV format and save the file with the same name as original time-lapse image file. Relevant data is later collated by the user

travelled data was collected and analyzed. Depending on research questions, the user may decide to collect one or more features from TrackMate output data.

3. Implementation of the TrackMate ImageJ plugin using this protocol will provide the end user with a high volume of lysosomal movement data, as the application of several filters will allow for the automated detection of numerous fluorescent signals. In using this protocol, we evaluated whether there was a difference in the distribution of total distance travelled by lysosomes depending on a specific treatment, so statistical analysis of the cumulative distribution was performed. The test recommended for this is the Kolmogorov–Smirnov t-test for

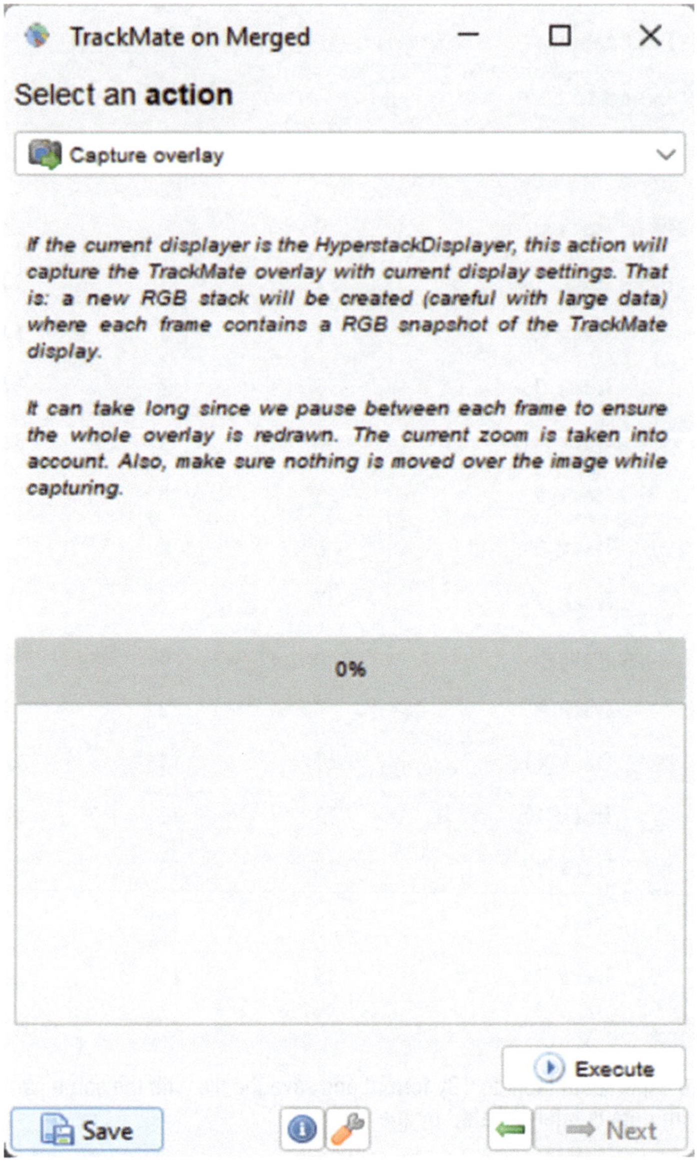

Fig. 11 Select "Capture Overlay" to generate a new image sequence with embedded spot and track overlays. Save the new image sequence file

cumulative distribution. Depending on the final application, the user may choose another analysis methodology.

4. While implementation of the TrackMate plugin in this manner provides a large volume of data on lysosomal movement, it did not provide information on directionality, anterograde and/or retrograde movement in relation to the cell soma. It is recommended that the application of this methodology be used in conjunction with kymographs [15, 16] to provide a more comprehensive view of trafficking data.

References

1. Settembre C, Zoncu R, Medina DL et al (2012) A lysosome-to-nucleus signalling mechanism senses and regulates the lysosome via mTOR and TFEB: self-regulation of the lysosome via mTOR and TFEB. EMBO J 31(5):1095–1108
2. Farfel-Becker T, Roney JC, Cheng XT et al (2019) Neuronal Soma-derived degradative lysosomes are continuously delivered to distal axons to maintain local degradation capacity. Cell Rep 28(1):51–64.e4
3. Ivanova D, Cousin MA (2022) Synaptic vesicle recycling and the Endolysosomal system: a reappraisal of form and function. Front Synaptic Neurosci 14:826098
4. Kulkarni VV, Anand A, Herr JB et al (2021) Synaptic activity controls autophagic vacuole motility and function in dendrites. J Cell Biol 220(6):e202002084
5. Roney JC, Li S, Farfel-Becker T et al (2021) Lipid-mediated motor-adaptor sequestration impairs axonal lysosome delivery leading to autophagic stress and dystrophy in Niemann-Pick type C. Dev Cell 56(10):1452–1468.e8
6. Zhang X, Wei M, Fan J et al (2021) Ischemia-induced upregulation of autophagy preludes dysfunctional lysosomal storage and associated synaptic impairments in neurons. Autophagy 17(6):1519–1542
7. Basak I, Hansen RA, Ward ME et al (2021) Deficiency of the lysosomal protein CLN5 alters lysosomal function and movement. Biomolecules 11(10):1412
8. Pesaola F, Quassollo G, Venier AC et al (2021) The neuronal ceroid lipofuscinosis-related protein CLN8 regulates endo-lysosomal dynamics and dendritic morphology. Biol Cell 113(10): 419–437
9. Yasa S, Sauvageau E, Modica G et al (2021) CLN5 and CLN3 function as a complex to regulate endolysosome function. Biochem J 478(12):2339–2357
10. Fernandopulle MS, Prestil R, Grunseich C et al (2018) Transcription factor-mediated differentiation of human iPSCs into neurons: rapid differentiation of iPSCs into neurons. Curr Protoc Cell Biol 79(1):e51
11. Wang C, Ward ME, Chen R et al (2017) Scalable production of iPSC-derived human neurons to identify tau-lowering compounds by high-content screening. Stem Cell Rep 9(4): 1221–1233
12. Schindelin J, Arganda-Carreras I, Frise E et al (2012) Fiji: an open-source platform for biological-image analysis. Nat Methods 9(7): 676–682
13. Ershov D, Phan MS, Pylvänäinen JW et al (2022) TrackMate 7: integrating state-of-the-art segmentation algorithms into tracking pipelines. Nat Methods 19(7):829–832
14. Tinevez JY, Perry N, Schindelin J et al (2017) TrackMate: an open and extensible platform for single-particle tracking. Methods 115:80–90
15. Jakobs MAH, Dimitracopoulos A, Franze K (2019) KymoButler, a deep learning software for automated kymograph analysis. eLife 8: e42288
16. Mangeol P, Prevo B, Peterman EJG (2016) KymographClear and KymographDirect: two tools for the automated quantitative analysis of molecular and cellular dynamics using kymographs. Mol Biol Cell 27(12):1948–1957

Chapter 12

Generation of Donor-Specific iPSC for Modelling Lysosomal Storage Disorders

Sueanne Chear, Adelene Chiam, Jana Talbot, Bryony N. Thorne, Emma J. Wilkinson, Alex W. Hewitt, and Anthony L. Cook

Abstract

iPSC technology has enabled the generation of human cell-based models of lysosomal storage disorders and has provided disease-relevant systems to undertake drug discovery or pre-clinical testing of gene- or cell-based therapies. Here, we provide a protocol to generate iPSCs derived from people with lysosomal storage disorders and illustrate expected results using a CLN2 disease donor-specific skin fibroblast culture. Protocol steps include lipofection of episomal plasmids, picking of putative iPSC colonies following live cell TRA-1-60 immunofluorescence, and quality control steps such as immunofluorescence for expression of undifferentiated cell markers, germ layer differentiation, and confirmation of pathological variant genotype. The iPSC generated by this protocol can be differentiated to several cell lineages and can be used with CRISPR/Cas technology to generate isogenic disease models.

Key words Neuronal ceroid lipofuscinosis, TPP1, Live cell immunofluorescence, Cell reprogramming

1 Introduction

Human cell-based models of disease, and specifically induced pluripotent stem cell (iPSC) models, are ideally suited to the study of genetic conditions such as lysosomal storage disorders because they can be generated in a donor-specific manner and then differentiated into affected cell types. iPSC technology has been used to generate new models of several lysosomal storage disorders (LSDs), including various forms of neuronal ceroid lipofuscinosis [1–5], Pompe disease [6], Fabry disease [7], multiple sulfatase deficiency [8], metachromatic leukodystrophy [9, 10], Niemann-Pick Type C disease [11, 12], various forms of mucopolysaccharidosis [13–16], Tay-Sachs disease [17], Sandhoff disease [18], globoid cell leukodystrophy [19], and Wolman disease [20]. The predominance of neurological phenotypes in LSDs has precipitated much

Stephanie M. Hughes and Indranil Basak (eds.), *Lysosomes: Methods and Protocols*, Methods in Molecular Biology, vol. 2976, https://doi.org/10.1007/978-1-0716-4844-5_12,

iPSC-based research into how disease-associated gene variants affect neurons and glia. However, the ubiquitous expression of lysosomal enzymes means that LSDs can affect diverse cell types, many of which can also be generated from iPSCs, including retinal photoreceptors [3], and cardiac muscle cells [6, 7]. A major strength of iPSC technology is the preservation of genetic backgrounds that drive cell type- and donor-specific patterns of gene expression, which has enabled in vitro modelling of disease-relevant genotype-phenotype correlations seen in vivo [21]. Most importantly, iPSC-based models have enabled drug discovery and pre-clinical testing of novel therapeutic strategies, including small-molecule pharmacological agents [20, 22, 23], enzyme replacement therapies [20], antisense oligonucleotides [24], gene therapies [25], and cell-based therapies [26, 27]. The power of iPSCs for providing new therapies is exemplified by recent work on amyotrophic lateral sclerosis, where iPSC-derived neurons were used for drug screening, ultimately identifying a lead compound that is now subject to clinical trial [23, 28]. With the US Congress via the *Food and Drug Administration Modernization Act 2.0* enabling alternatives to animal studies for supporting drug and biological agent applications, and the power of iPSCs to provide human cell-based models, it is likely such examples will become more prevalent.

The generation of iPSCs can be done by several methods. The original method developed for reprogramming fibroblasts to iPSCs used a retroviral delivery system that expressed four transcription factors: OCT4, SOX2, KLF4, and C-MYC [15, 16]. However, viral vectors can integrate into the cellular genome, carrying a risk for insertional mutation that may affect genome stability of iPSCs and promote tumorigenesis in relation to C-MYC oncogene expression [17]. Therefore, research with translational aspirations should aim to generate "footprint-free" iPSCs that are devoid of residual transgene sequences of the reprogramming vector. The integration-free methods involve the use of mRNA [18], Sendai virus [19, 20], and episomal plasmids [1, 21], which are readily available as reprogramming kits from commercial suppliers. mRNA reprogramming involves transfecting cells with in vitro-transcribed mRNAs that express specific transcription factors. The modified mRNAs do not integrate into the host genome, avoiding the need to screen for vector clearance after multiple passages. Although the mRNA-based method has a higher reprogramming efficiency (up to 2%) with a relatively short time to iPSC colony emergence, it is more costly and requires repeated mRNA transfection [29, 30]. Sendai virus is an RNA virus that replicates in the cytoplasm; therefore, there is no risk of integration in the host genome. Reprogramming efficiency has been reported to be 0.1% in CD34+ cells [24] and 1% in dermal fibroblasts [19]. Despite the high reprogramming efficiency, it takes up to 10 passages for the viral transcripts to be depleted from the host cytoplasm [25], and to eliminate viral particles, cells may need to be cultured at 39 °C [26].

In contrast, episomal vectors such as those based on oriP/EBNA1 (Epstein-Barr nuclear antigen-1) plasmid vectors are non-viral, safe to use, and inexpensive. The vectors described in this study are a combination of three episomal vectors with an oriP/EBNA1 backbone to deliver five reprogramming genes, OCT4, SOX2, LIN28, L-MYC, and KLF4 with high efficiency [31]. These are supplemented by two additional plasmids that express a p53 dominant negative mutant (mp53DD) and EBNA1 genes. Suppression of the p53 protein that is highly involved in tumor suppression through mp53DD mutation can increase reprogramming efficiency [32], while EBNA1 gene expression enables high expression of the reprogramming plasmids [33] but only transient expression of the EBNA1 plasmid due to lack of the oriP gene. This, along with the imperfect nature of oriP/EBNA1 partitioning, enables the removal of episomal vectors from the iPSCs at a rate of 2–8% per cell cycle [34]. Although reprogramming efficiency of dermal fibroblasts with episomal vectors is low, sufficient iPSC colonies can be generated from a reasonable number of starting cells with efficiency reported to be 0.0003–0.0006% in human foreskin fibroblast cultures [35]. The episomal-based method generates iPSCs from fibroblasts and blood cells more reliably with quicker loss of reprogramming vector through cell division and serial dilution in comparison to the Sendai virus-based method [31, 36].

Here, we describe a protocol to reprogram fibroblasts derived from donors with neuronal ceroid lipofuscinosis to iPSCs using episomal vectors. The protocol is based on our previously published iPSCs derived from a CLN3 donor fibroblast culture [1]. Here, we present data from the generation of an iPSC line using fibroblasts cultured from a donor harboring homozygous missense variants causing CLN2 disease (homozygous TPP1, c.509–1 G > C). CLN2 disease, one of 13 forms of neuronal ceroid lipofuscinosis, has an onset at 2–4 years of age and is characterized by seizures, vision loss, progressive motor and cognitive decline, and premature death [37, 38]. The iPSCs derived from these cultures exhibit pluripotency, lack genomic aberrations, retain donor-matched identity, and can be differentiated into a variety of cell types relevant to brain diseases, including neurons (Fig. 1, and references [1, 39]), and pericytes [40]. We have also successfully used these iPSCs to generate isogenic disease models by correcting the disease-associated gene variant [1, 41]. Thus, we anticipate the protocols herein will enable others to generate iPSC-based models of LSDs more broadly.

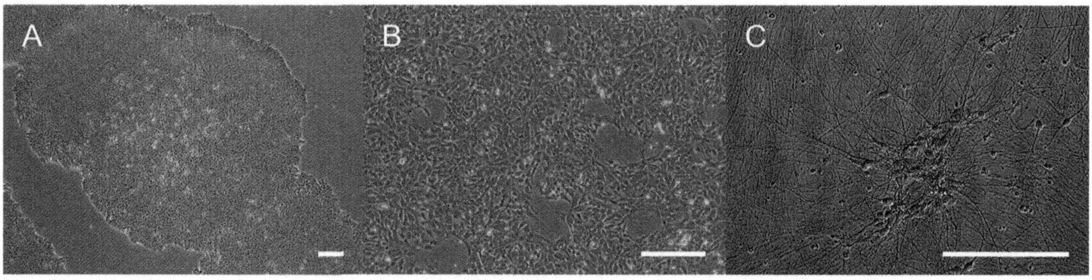

Fig. 1 Morphology of CLN2 iPSCs and their differentiation into neurons. Phase contrast images depict (**a**) iPSC culture at passage 14 with typical iPSC morphology and minimal differentiation; (**b**) neural stem cells at passage 2; and (**c**) maturing cortical neurons at DIV14. Scale bars: 200 μm

2 Materials

Cell Culture

Ethics and Biosafety Considerations

Obtaining and working with human samples requires approval from relevant ethics bodies, and the generation of iPSCs requires biosafety approvals; researchers should ensure their own work conforms with local requirements. The work presented herein was performed under Human Research Ethics Committee-Tasmania approval (H26563) and approval from the University of Tasmania Institutional Biosafety Committee.

General Considerations

All procedures involving medium preparation, medium exchange, or handling of cells should be performed in class II biological safety cabinets to maintain sterility. Cells should be regularly inspected for microbial contamination. Aspects that should be periodically documented using routine microscopy include cell and/or colony morphology as well as development, changes in media color, as well as any abnormal observations (e.g., excessive cell death, unexpected halt in proliferation). Routine testing for mycoplasma should also be conducted to exclude contamination.

2.1 Fibroblast Cells

Fibroblast cultures should be free from microbial contamination and have low passage number, ideally passage 2 or 3 (P2–P3), although slightly higher passage numbers (P4–P6) can also be used. Many academic repositories have extensive collections of patient-specific fibroblast cultures derived from several LSDs (e.g., Coriell repository). Alternatively, fibroblast cultures can be established from skin biopsy using protocols described elsewhere [42]. Other cell types, such as peripheral blood mononuclear cells, can also be used in reprogramming experiments to generate iPSCs [43].

2.2 Media and Other Culture Reagents

1. 10 μg/mL Vitronectin XF.
2. Matrigel hESC-Qualified Matrix, LDEV-free.
3. Complete fibroblast medium: 88% Dulbecco's Modified Eagle Medium/F12 (DMEM) with high glucose supplemented with 10% fetal bovine serum (FBS), 1% MEM Non-Essential Amino Acids Solution, and 1% Antibiotic-Antimycotic.
4. Dulbecco's phosphate-buffered saline, no calcium no magnesium (DPBS).
5. 0.25% Trypsin-EDTA.
6. Fibroblast cryopreservation medium: 10% dimethyl sulfoxide (DMSO) and 90% FBS.
7. Opti-MEM 1 Reduced Serum medium.
8. N2B27 medium: 97% DMEM/F-12 with HEPES, 1% N-2 supplement, 1% B-27 supplement, 1% MEM Non-Essential Amino Acids Solution, 0.1 mM 2-mercaptoethanol supplemented with 100 ng/mL human recombinant basic fibroblast growth factor (bFGF) (*see* **Note 1**).
9. Essential 8 medium.
10. ReLeSR.
11. iPSC cryopreservation medium: 90% KnockOut Serum Replacement (KSR), 10% DMSO.
12. Anti-Adherence Rinsing Solution.
13. DMEM/F12 with HEPES.
14. Complete KSR-EB medium: 20% KSR, 79% DMEM/F-12 with GlutaMAX supplement, 1% MEM Non-Essential Amino Acids, 0.1 mM 2-mercaptoethanol.

2.3 Molecular Reagents

1. Epi5 Episomal iPSC Reprogramming Kit: Epi5 Reprogramming Vectors, Epi5 p53 and EBNA vectors.
2. Lipofectamine 3000 Transfection Reagent: Lipofectamine 3000 Reagent, P3000 Enhancer Reagent.
3. EBNAl primers (*see* Table 1).
4. TaqMan assay probes (*see* Table 2).
5. HumanCytoSNP-12 BeadChip array.

Table 1
EBNA1 primers used to confirm removal of episomal vectors

Oligo name	Sequence	Expected PCR band size
EBNA-1-F	ATC GTC AAA GCT GCA CAC AG	666 bp
EBNA-1-R	CCC AGG AGT CCC AGT AGT CA	

Table 2
List of TaqMan assay probes for iPSC differentiation

Marker	Gene name	TaqMan assay ID
Undifferentiated state	*OCT4*	Hs01895061_ul
Undifferentiated state	*NANOG*	Hs00415716_ml
Undifferentiated state	*SOX2*	Hs00415716_ml
Ectoderm	*NEFH*	Hs00606024_ml
Ectoderm	*PAX6*	Hs01088114_ml
Mesoderm	*HAND2*	Hs00232769_ml
Mesoderm	*ESM1*	Hs00199831_ml
Endoderm	*AFP*	HHs00173490_ml
Endoderm	*SST*	Hs00356144_ml
Endogenous control	*HPRT1*	Hs99999909_ml
Endogenous control	*EEF2*	Hs00157330_ml
Endogenous control	*RPLP0*	Hs04189669_gl

Table 3
CLN2 primers used to confirm presence of SNV (c.509-1G>C)

	Sequence	Expected PCR band size
Forward (FWD)	GGA GAC AGT GAC TGA CTG CCC AGG	721 bp
Reverse (REV)	GTT GCC ACC GAA GAG GCG CAT GAA C	

6. Primers to amplify region of pathogenic gene variant (*see* Table 3 *for CLN2-specific example*).
7. Tris-acetate-EDTA (TAE) Buffer, Molecular Biology Grade.
8. SYBR Safe DNA Gel Stain.
9. *Optional:* RNeasy Plus Mini Kit.
10. *Optional:* QIAamp DNA Mini kit.
11. *Optional:* Omniscript RT kit.
12. *Optional:* TaqMan Gene Expression Master Mix.
13. *Optional:* TaqMan hPSC Scorecard Kit.

2.4 Immuno-cytochemistry-Immunofluorescence

1. TRA-1-60 Antibody, anti-human, REAfinity Live cell stain.
2. 1× Hanks' Balanced Salt Solution (HBSS).
3. 4% paraformaldehyde in HBSS.

4. Blocking solution: 5% FBS and 0.1% Triton X-100 in HBSS for intranuclear markers or 5% FBS in HBSS for cell-surface markers.
5. StemLight Pluripotency Antibody Kit:
   (a) Intranuclear markers: rabbit anti-NANOG (1:200), rabbit anti-OCT4A (1:200), rabbit anti-SOX2 (1:200).
   (b) Cell-surface markers: mouse anti TRA-1-60(S) (1:200), mouse anti- TRA-1-81 (1:200), mouse anti-SSEA4 (1:200).
6. Secondary antibodies: goat anti-mouse IgG, goat anti-rabbit IgG.
7. 1 μg/mL DAPI in HBSS.
8. 0.02% Sodium azide in HBSS.

3 Methods

3.1 Fibroblast Culture

1. Culture fibroblasts in complete fibroblast medium, replenishing medium every 3 days until ~80% confluency.
2. Fibroblasts can be passaged using Trypsin-EDTA, and cryopreserved (*see* **Note 2**), or re-plated for further growth/expansion or for generation of iPSCs.
3. When fibroblast cultures are ~80% confluent, collect cells using Trypsin-EDTA, count using a haemocytometer or automated cell counter, and plate 70,000 cells/well on a non-tissue culture-treated 6-well plate coated with Vitronectin XF according to the manufacturer instructions.
4. Incubate cells at 37 °C and 5% CO_2 overnight.
5. The following morning, wash cells with DPBS (2 mL/well) and replace medium with antibiotic- and antimycotic-free fibroblast medium. Return cells to incubator while preparing liposomes.

3.2 Lipofectamine Transfection of Fibroblasts

1. Pre-warm OptiMEM 1 Reduced Serum medium to room temperature (RT).
2. For each well of a 6-well plate, prepare reprogramming vectors by aliquoting 118 μL pre-warmed Opti-MEM into a 1.5 mL microcentrifuge tube and adding 1.2 μL of Epi5 Reprogramming Vectors, 1.2 μL of Epi5 p53 and EBNA vectors, and 4.8 μL of P3000 reagent.
3. For each well of a 6-well plate, prepare liposomes by aliquoting 121 μL Opti-MEM into a 1.5 mL microcentrifuge tube and adding 3.6 μL of Lipofectamine 3000 reagent. Mix gently.

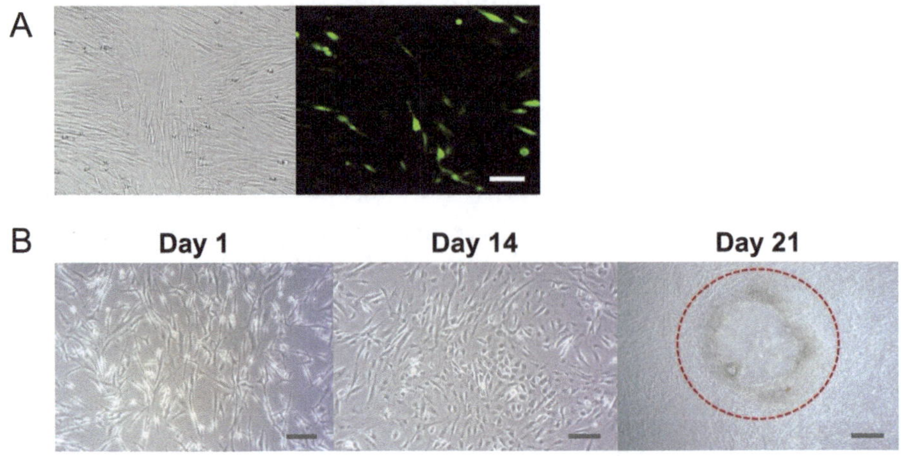

Fig. 2 Reprogramming of human fibroblasts to iPSCs. (**a**) Lipofection control cells demonstrating GFP-positive fibroblasts under bright-field (left panel) and fluorescence microscopy (right panel). (**b**) Morphological transition of fibroblasts to iPSCs as observed on day 1, 14, and 21 post-lipofection. Putative iPSC colonies on day 21 are highlighted by red, dashed outlines. Scale bars: 100 μm

4. Carefully combine and mix the contents of the two tubes to produce a transfection master mix and incubate at RT for 5–15 min (*see* **Note 3**).
5. Add the transfection mix to the fibroblast cultures, return the cells to the incubator, and culture for 24 h (*see* **Note 4**).
6. After 24 h, replace the fibroblast medium containing liposomes with 2 mL N2B27 medium, with freshly added 100 ng/mL bFGF (*see* **Note 1**).
7. Continue to culture the transfected fibroblasts for a total of 14 days, replenishing the culture with 2 mL N2B27 medium daily.
8. On day 14, replace the N2B27 medium with 2 mL complete Essential 8 medium, and replenish Essential 8 medium every day thereafter to day 21.
9. iPSC colonies of approximately 200 μm will emerge among the now highly confluent fibroblasts and can be visually recognized by their typical iPSC colony morphology (Fig. 2).

3.3 Live-Cell Immunostaining of Reprogrammed Fibroblasts

1. On day 21, prepare a live-cell staining solution by diluting TRA-1-60 antibody (1:100) in fresh Essential 8 medium (*see* **Note 5**). About 1–2 mL of staining solution is sufficient for 1 well of a 6-well plate.
2. Aspirate the media from the well and gently wash once with DPBS. Add the staining solution to each well and gently rock the plate to evenly distribute the solution. Incubate at 37 °C and 5% CO_2 for 30 min.

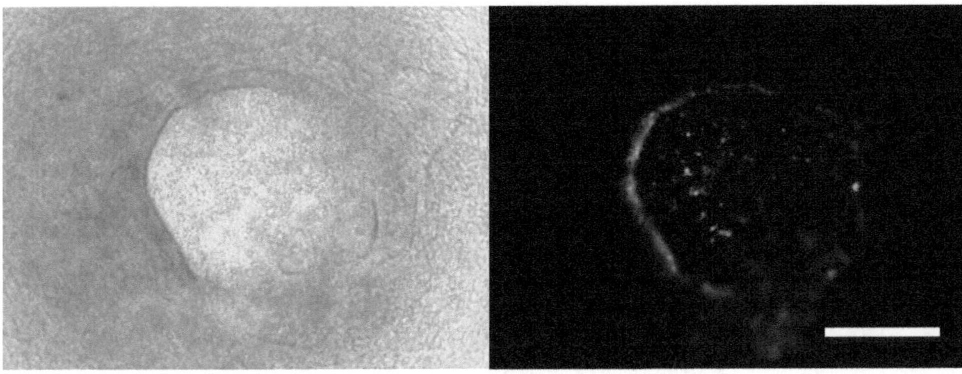

Fig. 3 Live-cell TRA-1-60 immunofluorescence. Live-cell immunofluorescence was carried out on reprogrammed CLN2 fibroblast culture 21 days after lipofection to identify undifferentiated cells through TRA-1-60 antibody staining. Representative phase contrast (left panel) and fluorescence (right panel) images of TRA-1-60 staining showed putative iPSC colonies. Scale bar: 200 μm

3. Aspirate the staining solution and wash cells twice with Essential 8 medium to remove any excess antibody.
4. Add fresh Essential 8 medium to each well. Examine the cells under a fluorescence microscope. Putative iPSC colonies will stain positively for the marker of undifferentiated state, TRA-1-60, indicating successful reprogramming of the fibroblasts (Fig. 3).

3.4 Monoclonal Expansion of iPSCs

1. To isolate the iPSC colonies, manually pick them under a stereomicroscope (*see* **Note 6**).
2. Cut each TRA-1-60 positive colony into 10–20 pieces in a grid-like pattern using a sterile 29-gauge needle.
3. Using a P200 micropipette, gently transfer the cut pieces into each well of a Vitronectin XF-coated 12-well plate containing Essential 8 medium for further expansion (*see* **Note 7**).
4. Inspect the iPSCs daily for any spontaneous differentiation. If present, manually remove the differentiated cells using the sterile tip of a P200 micropipette under a stereomicroscope starting from ~4 days post-passage (*see* **Notes 8** and **9**).
5. Once the iPSCs have achieved ~80% confluency, they are ready for passage. Harvest the iPSCs using ReLeSR according to manufacturer instructions and replate cell aggregates into prepared wells of 6-well plates at a ratio of 1:20–1:50 (*see* **Note 10**).
6. Culture the iPSCs in 6-well plates for further expansion, cryopreservation, and/or validation assays.

3.5 Cryopreservation of iPSCs

1. To cryopreserve iPSC, harvest cells as described for passaging and resuspend cell aggregates in pre-chilled iPSC cryopreservation medium.
2. Using a 5 mL serological pipette, aliquot 1 mL into each cryovial (*see* **Note 11**), taking care not to break up the iPSC aggregates. Do not generate a single-cell suspension.
3. Place the vials in a suitable cryopreservation container (e.g., CoolCell LX freezing container), which can facilitate controlled, slow-rate cooling (i.e., −1 °C/min).
4. Immediately transfer the container to −80 °C.
5. Once the cells have equilibrated to −80 °C, transfer the vials to long-term storage at < −170 °C (liquid nitrogen) (*see* **Note 12**).

3.6 Detection of Episomal Reprogramming Vectors

From passage 2 (P2), newly generated iPSCs should be assessed for the presence of episomal reprogramming vectors. This can be done using "end-point" PCR using DNA isolated at each passage.

1. Extract DNA using preferred method (*see* **Note 13**).
2. Set-up PCR using primers EBNA-1-F and EBNA-1-R (Table 1), which detect all 5 episomal plasmids, and amplify DNA using cycling conditions of 95 °C for 3 min, followed by 30 cycles of 98 °C/20 s, 63 °C/15 s, 72 °C/60 s, with a final extension step of 72 °C for 60 s.
3. Electrophorese PCR products on a 1% agarose gel in 1× Tris-acetate-EDTA (TAE) buffer, using SYBR Safe DNA Gel Stain to visualize the amplicons. The expected PCR product size is 666 bp. Typically episomal reprogramming vectors are not detectable by ~ P8 (Fig. 4).

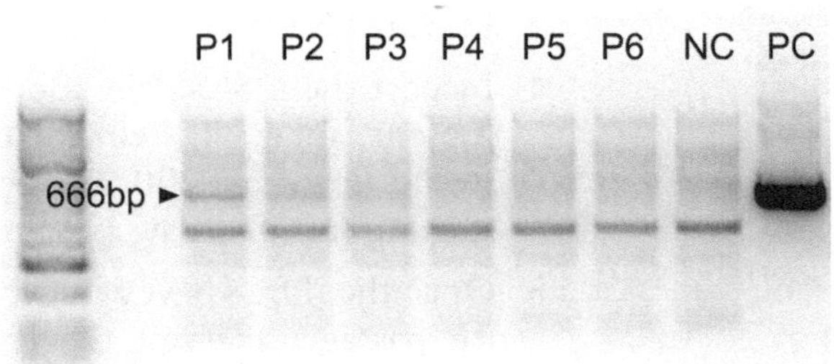

Fig. 4 Detection of EBNA1 episomal vector. End-point PCR showed that EBNA1 vector was completely absent in CLN2 iPSCs after ~4 passages. NC, negative control from non-transfected fibroblasts; PC, positive control from EBNA1 plasmids

3.7 Immunofluorescence for Markers of the Undifferentiated State

1. Harvest iPSCs using the standard passaging protocol as described above (*see* Subheading 3.4).
2. Plate cell aggregates onto Matrigel-coated 96-well black flat bottom tissue culture-treated microplates (*see* **Note 14**), and culture until ~60% confluency is reached.
3. Fix cells with 4% paraformaldehyde for 10–15 min at RT on an orbital shaker at ~50 rpm. Unless otherwise specified, allow the plate to sit on the shaker during all incubation steps and in-between washes.
4. Wash the cells three times in HBSS for 10 min per wash.
5. Apply the appropriate blocking solution (*see* **Note 15**) to each well and incubate for 30 min at RT. During this time, prepare the primary antibody solution by diluting the antibody in HBSS.
6. Remove the blocking solution. Without washing the cells, immediately add the primary antibody solution to the wells, using 50 μL per well. Incubate overnight at 4 °C.
7. Remove the antibody solution and wash the cells as described in **step 4**.
8. Prepare the secondary antibody solution similarly by diluting the antibodies in HBSS. Add the solution to the wells and incubate for 2 h at RT (*see* **Note 16**).
9. Remove the antibody solution and immediately add DAPI. Incubate for 15 min at RT.
10. Wash the cells as described in **step 4** and add HBSS with 0.02% sodium azide. Image cells using standard fluorescence microscopy (*see* **Note 17**). See Fig. 5 for expected results.

3.8 Virtual Karyotyping

There are multiple technologies that can be used for karyotyping of iPSCs. Virtual karyotyping using genome-wide SNP arrays to quantify copy-number variation (CNV) can be performed "in house" or via the use of third-party providers. We have used Australian Genome Research Facility (Melbourne, Australia) for CNV analysis using the HumanCytoSNP-12 beadchip array, using 200 ng of purified genomic DNA. This generates data for over 300,000 SNPs across the genome with a median spacing of ~6 kb and can be used to detect small deletions and amplifications, a higher resolution output relative to Giemsa staining (G-banding) which is in the order of 5–10 MBs. However, virtual karyotyping is less suitable for the detection of balanced chromosomal translocations which do not involve copy-number variations. Due to existing detailed protocols for analysis of SNP array data [44], we instead provide a concise overview of how we analyze and interpret virtual karyotyping output, using example data from our CLN2 iPSCs presented here.

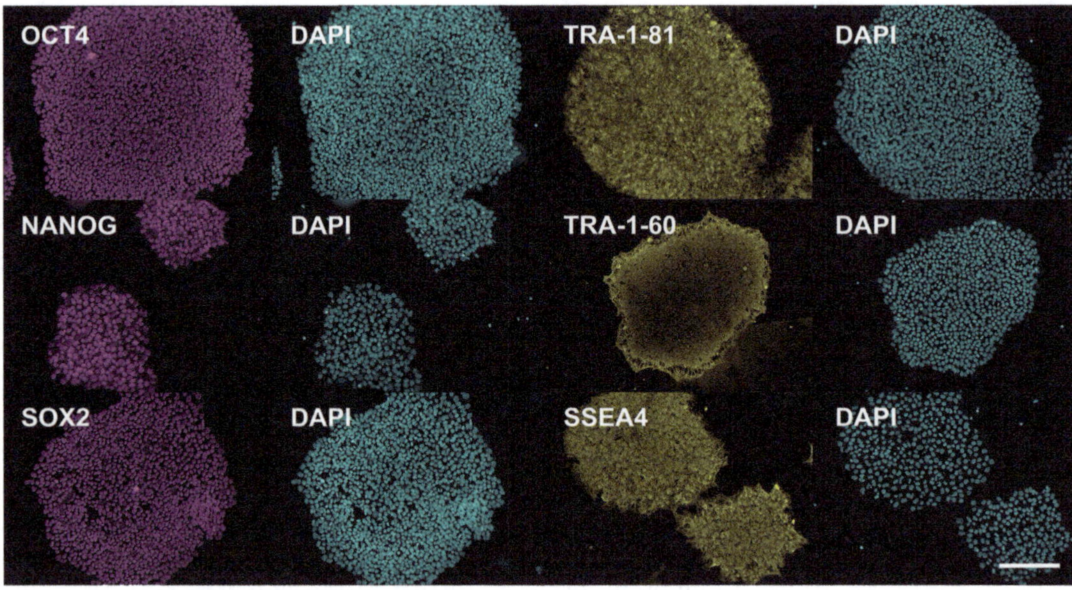

Fig. 5 Immunofluorescence with markers of the undifferentiated state. CLN2 iPSCs displayed nuclear (OCT4, NANOG, and SOX2) and cell-surface (TRA-1-60, TRA-1-81, and SSEA-4) markers indicating their undifferentiated developmental state. Scale bar: 200 μm

1. Extract genomic DNA from iPSCs, using cultures at P12–P15.
2. Determine the B allele frequency (BAF; *see* **Note 18**) and log R ratio (LRR; *see* **Note 19**) of each single nucleotide polymorphism (SNP) marker using GenomeStudio.
3. Analyze these parameters using PennCNV [45] and QuantiSNP [46] with default parameter settings. Identify genomic regions with CNV calls having at least 20 contiguous SNPs or genomic regions with SNPs spanning at least 1 MB.
4. Visualize these regions of interest using GenomeStudio and confirm the absence of chromosomal aberrations [42]. See Fig. 6 for representative data from CLN2 fibroblasts and iPSCs which collectively demonstrate a normal male karyotype [46, 47] without reprogramming-induced chromosomal aberrations.

3.9 Embryoid Body Formation and Germ Layer Differentiation Assay

Germ layer differentiation in iPSC-derived embryoid bodies (EBs) is a well-established functional assay for the validation of pluripotency in iPSCs. The efficiency of iPSC differentiation can be assessed via quantitative gene expression analysis to generate reproducible and comparable datasets. To ensure that findings can be interpreted meaningfully, it is imperative that all three germ layers (i.e., ectoderm, mesoderm, endoderm) are considered. We quantitate the expression levels of at least two markers corresponding to each germ layer in addition to three markers of the undifferentiated

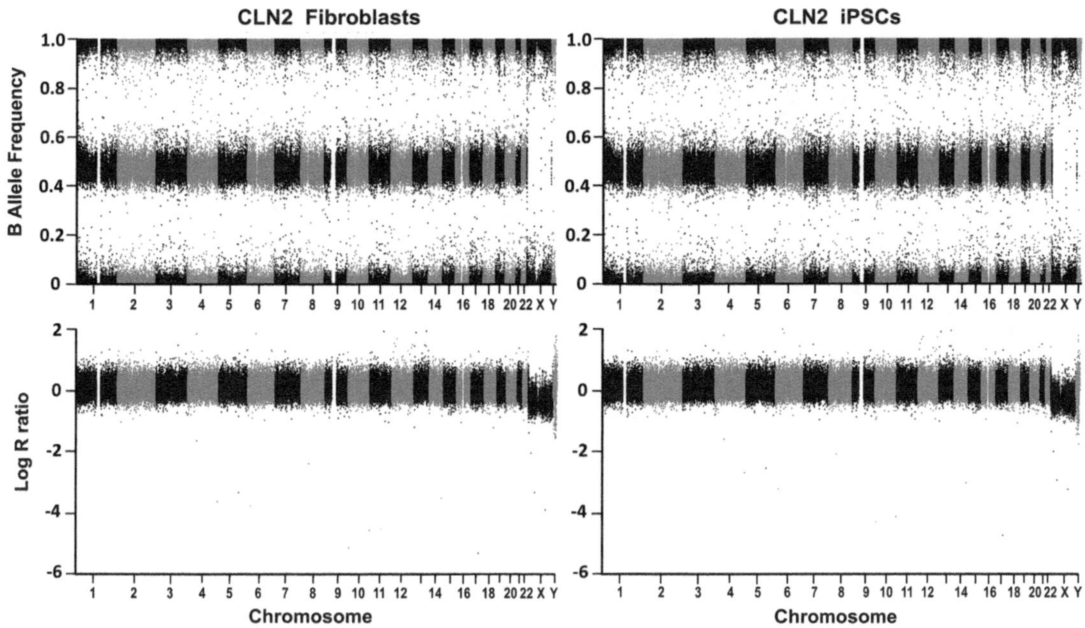

Fig. 6 Copy number variation analysis of CLN2 iPSCs. Panels show BAF (top panels) and LRR (bottom panels), respectively, for the CLN2 parental fibroblasts (left panels) and CLN2 iPSCs (right panels)

state (*see* Table 2), assessing changes across different assay time points. Here, we describe a detailed method for the generation and maintenance of EBs, with brief recommendations for quantitative gene expression analysis.

1. Expand iPSCs in a 6-well plate and culture until they reach ~80% confluency (*see* **Note 20**).
2. Aliquot a sufficient volume of KSR-EB medium (*see* **Note 21**). Pre-warm to RT.
3. Add 1 mL of KSR-EB medium to each well of an ultra-low attachment plate that will receive cells (*see* **Note 22**).
4. Wash the cells using DPBS. Use one well for RNA extraction (DIV 0, control time point) (*see* **Note 23**).
5. For the remaining wells, passage cells using ReLeSR and collect cell aggregates in KSR-EB medium, transferring them gently to a 15 mL conical tube using a 5 mL serological pipette.
6. Centrifuge cells at 100× *g* for 3 min at RT.
7. Aspirate the supernatant and gently flick the bottom of the tube to dislodge the cell pellet. Resuspend in 2–4 mL of KSR-EB medium, taking care to preserve the integrity of the aggregates as much as possible.
8. Repeat **steps 5** and **6** (for a total of two washes).
9. Pool cells from two wells of a 6-well plate into one well with KSR-EB medium (*see* **Note 24**). Incubate for 48 h at 37 °C.

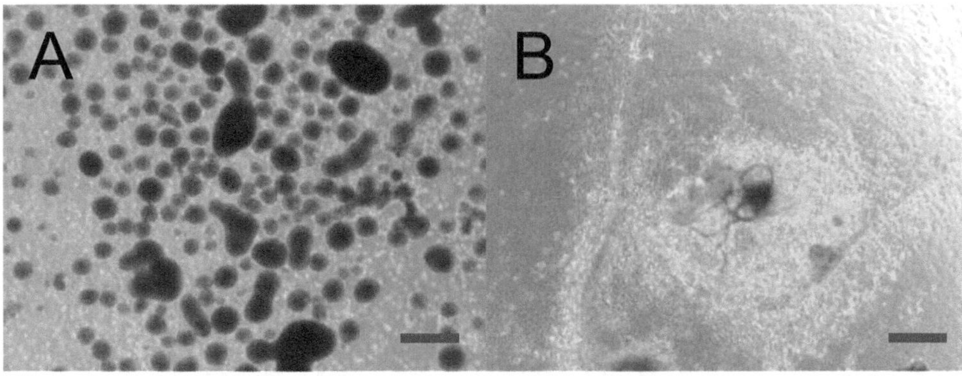

Fig. 7 Embryoid body (EB) formation and spontaneous differentiation. (**a**) EBs formed from iPSC aggregates and cultured in suspension showed smooth round borders on DIV 2. (**b**) Upon attachment to culture wells, EBs began to differentiate into diverse types of cells as seen on DIV 14. Scale bar: 100 μm

10. On DIV 2, perform a media change. Transfer the cell suspension of each well into a conical tube and allow EBs to gradually settle to the bottom. Gently remove the supernatant using a serological pipette (not a vacuum-fitted aspiration). Resuspend EBs in fresh KSR-EB medium pre-warmed to 37 °C and transfer them back to the ultra-low attachment plate. Incubate for another 48 h at 37 °C. See Fig. 7 for representative DIV 2 EBs cultured in suspension.
11. On DIV 4, repeat Step 9 but pool all EBs together. After resuspending in fresh KSR-EB medium, split the EBs equally into separate wells on a Matrigel-coated tissue culture-treated 6-well plate at a ratio of 1:2 (*see* **Note 25**). Return the plate to the incubator and allow the EBs to attach.
12. For the first 3–4 days, perform routine media changes every other day. Start by using 3–4 mL of medium per well and carefully monitor cell growth. As cell proliferation increases with noticeable signs of germ layer differentiation (~DIV5 onwards), the volume of media can be increased to a maximum of 8 mL per well, and media changes can be done daily (*see* **Note 26**).
13. On DIV 14, there should be a range of cell morphologies suggestive of differentiation to multiple cell types. See Fig. 7 for representative images.
14. Aspirate the spent medium and wash the cells gently using DPBS.
15. Extract RNA from all wells. Quantify RNA concentration and purity (i.e., A260:A280 ratio). Store RNA samples at −80 °C.
16. Reverse transcribe 1 μg of RNA to generate cDNA (*see* **Note 23**). Prepare 10 ng/μL of cDNA (e.g., by diluting the reverse transcription reaction using nuclease-free water) and store at −80 °C.

17. Set up 5 μL qPCR mixes by combining 1 μL of the diluted cDNA with the master mix and primers/assay probes (*see* **Note 27**). Refer to Table 2 for target markers and their corresponding TaqMan probes.
18. Load the reaction mixes into 384-well clear reaction plates. Seal with optical film.
19. Load the plate onto an analyzer (e.g., QuantStudio 7) and initiate a qPCR using cycling conditions of 50 °C/2 min, then 95 °C/10 min, followed by 40 cycles of 95 °C/15 s, 60 °C/60 s.
20. Examine the qPCR data according to routine in-house methods (*see* **Note 28**) to determine the efficiency of germ layer differentiation. Undifferentiated marker genes should be substantially downregulated, whereas the expression of lineage markers for each differentiated germ layer should increase.
21. If desired, iPSC clones assessed by RT-qPCR can be further assessed using larger panels of lineage markers, such as the TaqMan hPSC Scorecard Panel. Data analysis can be easily performed using web-based software (*see* Fig. 8 and **Note 29**). Alternatively, immunofluorescence could be used to detect various lineages within the culture.

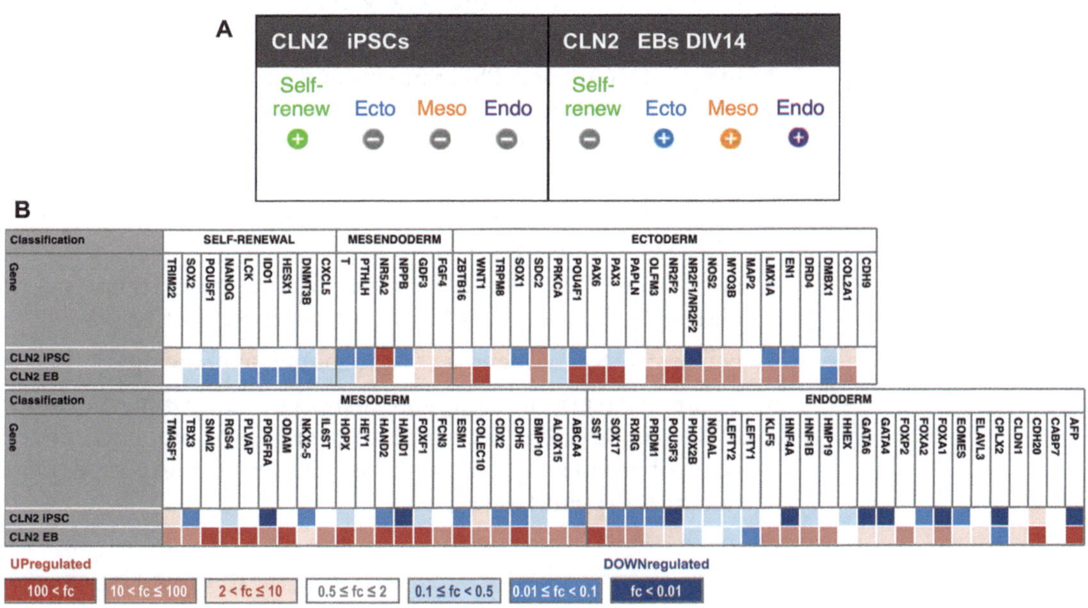

Fig. 8 Taqman hPSC scorecard analysis. (**a**) Overview of the analysis showed that CLN2 iPSCs were undifferentiated at starting point (i.e., DIV 0) and expressed markers for self-renewal. Conversely, after 14 days of spontaneous differentiation, EBs derived from these CLN2 iPSCs expressed markers for all three germ layers. (**b**) Heatmap shows changes in expression levels of self-renewal and germ layer-specific genes. Clear correlation between the differentiation state and relevant markers is visible

3.10 Confirmation of iPSC Identity Corresponding to Donor

Confirmation of cell line identity is a fundamental component of iPSC quality control. One of the most straightforward approaches to achieve this is by short tandem repeat (STR) analysis. However, it is important to note that STR raw data should not be published. Instead, the output should only be reported as the number of markers that are positive matches between the parental cell type (e.g., fibroblast) and iPSC (*see* **Note 30**).

Another aspect of iPSC identity that must not be neglected is the retention of any pathogenic gene variants originating from the donor. These may constitute single nucleotide changes or small deletions, such as the 966 bp deletion found in CLN3 disease [1]. The presence of these pathogenic gene variants can be confirmed in the iPSCs using PCR and size resolution following gel electrophoresis, as well as—if appropriate—sequencing of the PCR product. Here, we present a method to identify the known pathogenic single nucleotide variant (SNV) associated with CLN2 disease, i.e., c.509-1G > C. Nevertheless, the vast number of genes and pathogenic variants within each gene that cause LSDs will necessitate optimization of genotyping in a donor- and disease-specific manner. For instance, there are >150 known pathological variants in CLN2 alone, while there are >500 variants collectively in the 13 genes known to cause neuronal ceroid lipofuscinosis [38].

1. Identify the locus of the gene variant and design suitable primers that can amplify the region of interest.
2. Extract DNA from iPSC samples and perform end-point PCR on the DNA sample using the designed primers (*see* Table 3) and cycling conditions of 98 °C/30 s followed by 35 cycles of 98 °C/10 s, 72 °C/60 s, and a final extension step of 72 °C for 2 min. The expected PCR product size is 721 bp and can be resolved using gel electrophoresis.
3. Purify the DNA sample or perform a gel extraction (*see* **Note 31**). Measure DNA concentration and purity.
4. Perform DNA sequencing (e.g., Sanger sequencing).
5. Conduct a pairwise alignment analysis by comparing the donor iPSC sample DNA sequence to a reference sequence of the gene of interest and determine whether the expected gene variant is present (*see* Fig. 9).

4 Notes

1. Prepare the required volume of medium and add bFGF fresh prior to each use.
2. To cryopreserve fibroblasts, collect cells and pellet at 300× *g* for 3 min, resuspend in 10% DMSO in FBS, and aliquot ~250,000

CLN2 Reference sequence: NG_008653.1

…*ggagacagtgactgactgcccagg*gatgctcagaggtagcttcttctg
ttccgttttgagctttctgacctctgttctctgacctccagacaagcag
agctgctgctccctggggctgagtttcatcactatgtgggaggacctac
ggaaacccatgttgtaaggtccccacatccctaccagcttccacaggcc
ttggcccccatgtggactttggtaacacctatggggtgaatggggtt
ggggcacacagatccaggggctgaggaagtttagatgccattggggact
gggggtggggtgagttgtaaggtgggcattacagtctataagatctcct
caagcctgacttctccctaca**g**tgggggactgcaccgtttcccccaa
catcatccctgaggcaacgtcctgagccgcaggtgacaggactgtagg
cctgcatctgggggtaacccctctgtgatccgtaagcgatacaacttg
acctcacaagacgtgggctctggcaccagcaataacagccaagcctgtg
cccaggtgagccaagcaaagagccccagggtcctcatagcctccccaca
gtgtcctcaattccttaccaccctgggactcaccctcggacccacgatc
tctgctctgactccctccatagttcctggagcagtatttccatgactca
gacctggctca*gttcatgcgcctcttcggtggcaac*…

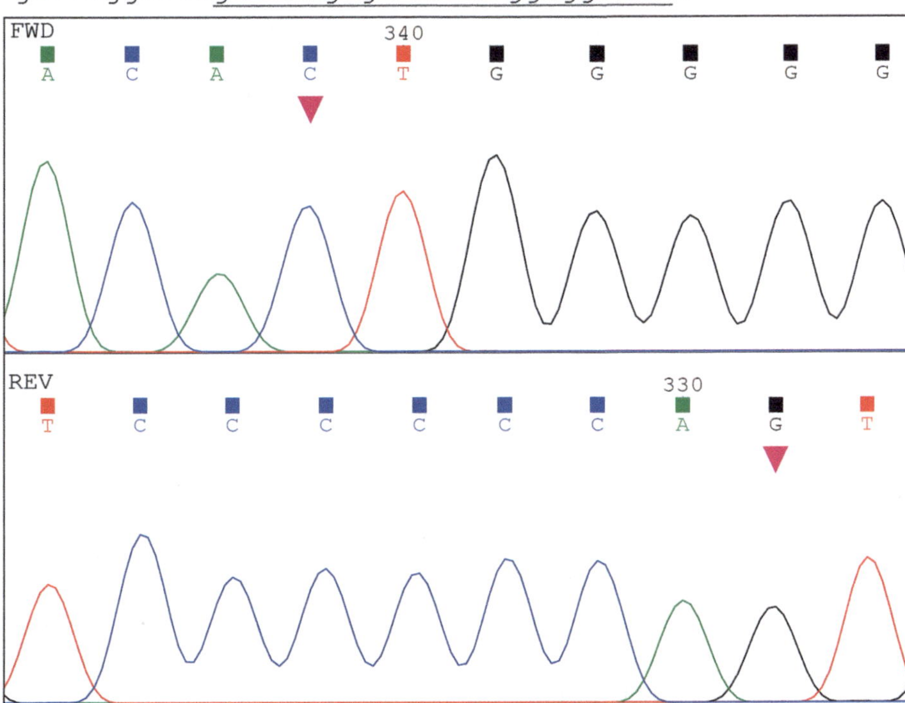

Fig. 9 DNA sequencing results for CLN2 iPSC showing homozygous c.509-1G>C. The nucleotide base highlighted in magenta represents the locus for a known pathogenic SNV associated with CLN2 disease, in which guanine (G) is substituted by cytosine (C). The short sequences underlined in italic are complementary to the primers used to verify the presence of the SNV via sequencing. DNA chromatograms of amplicons derived from the Forward (FWD) and Reverse (REV) primers both, respectively, show the G>C substitution (magenta arrows), indicating a homozygous mutation

cells/mL into cryotubes. Freeze as per for iPSCs and transfer to liquid nitrogen.

3. If desired, a plasmid expressing GFP or other readily visualized reporter gene can be transfected in a parallel well to confirm successful lipofection (*see* Fig. 2a).
4. An increase in cell debris suggestive of cell death may be observed, but this should become minimal by day 7 after transfection.
5. While we have found that an antibody dilution of 1:100 works best for direct live-cell staining, we recommend testing a range of dilutions (e.g., 1:50–1:200) to determine the optimal dilution prior to actual application. Protect the antibody from light.
6. The stereomicroscope should be stationed within a biological safety cabinet and fitted with a heated stage. Allow the heated stage to warm up to a temperature of 37 °C prior to use. The plate should not remain outside of an incubator for more than ~15 min to avoid any deleterious effects on cell viability. If more handling time is required, return the plate to the incubator and allow the cells to equilibrate for ~30 min before taking the plate out again.
7. Take care to maintain the integrity of the individual cut pieces and do not generate a single-cell suspension. A small cell lifter (e.g., double-ended cell lifter) may also be used for gentler cell handling.
8. We have noted that spontaneous differentiation was visibly decreased when transitioning the plate coating matrix from Vitronectin XF to Matrigel. This may vary depending on cell line and culture conditions. Nevertheless, iPSC cultures must be diligently monitored and differentiated cells should be removed as thoroughly as possible prior to each new passage and/or cryopreservation. This enables the reduction of spontaneous differentiation to as low as <1%, thus preserving the purity and homogeneity of the iPSC culture.
9. iPSC can be transitioned to other suitable media, such as mTeSR Plus following manufacturer instructions. We have found that transitioning to mTeSR Plus medium at P2 reduced spontaneous differentiation using this protocol.
10. Split ratios for the passaging of iPSCs represent the volume of resuspended aggregates that should be seeded. For example, when harvested iPSC aggregates are resuspended in 1 mL of medium, a 1:20 split ratio indicates that 50 μL of the aggregate suspension should be seeded into the new well.
11. We have found that at 70–80% confluency, iPSCs from 1 well of a 6-well plate can be split into 2–4 vials during

cryopreservation (each vial always contains 1 mL of cryopreservation media). This may vary with cell line, and hence, we suggest testing a range of split ratios.

12. When cryopreservation is performed quickly and correctly, iPSCs may be stored at −80 °C for up to 60 h with minimal impact on cell survival. We do not recommend keeping the iPSCs at this temperature for more than 3 days.
13. We prefer using the QIAamp DNA Kit.
14. We prefer to use specialized imaging microplates. Alternatively, iPSCs can also be cultured on sterile glass coverslips, pre-etched using nitric acid.
15. Typically, we use 5% FBS in 0.1% Triton X-100 and 1× HBSS as the blocking solution for routine immunocytochemistry-immunofluorescence. For cell-surface markers specifically (i.e., TRA-1-60(S), TRA-1-81, SSEA2), we instead opt for 5% FBS in 1× HBSS as per recommendations from the antibody manufacturer.
16. From this point onwards, protect from light.
17. Standard fluorescence microscopy typically provides sufficient resolution at 10–20× magnification to verify the expression of pluripotency markers. We use a Celldiscoverer 7 for visualization and imaging.
18. BAF is an estimate of the non-reference allele frequency and can be used to define regions of CNVs. See example plot in Wang et al. [45] for guidance in interpreting BAF plots for CNV-neutral loss-of-heterozygosity, 3- and 4-copy CNVs, relative to results expected for normal chromosomal regions.
19. LRR is the normalized signal intensity for each SNP. A ratio of 0 indicates two alleles are present, with higher and lower values indicating duplication or deletion, respectively.
20. We typically culture three wells of iPSC as a minimum to obtain RNA samples at DIV 0 (i.e., one well of starting iPSCs) and DIV 14 (i.e., two wells of iPSCs used for generating EBs). Additional time points or cryopreservation/continuation of the culture will necessitate additional wells.
21. KSR-EB medium may be supplemented with 5 μM ROCK inhibitor (Y-27632) to promote cell survival 24 h after plating into an ultra-low attachment plate.
22. Alternatively, create an ultra-low attachment plate by rinsing each well of a tissue-culture treated 6-well plate with 1 mL of Anti-Adherence Rinsing Solution, followed by two washes with DMEM/F-12 with HEPES.
23. RNA samples can be obtained using a preferred extraction method of choice, as long as the method is kept consistent

for the entire assay. We prefer using the RNeasy Plus Mini Kit as it allows us to obtain a sufficient yield of high-quality RNA that is compatible with the Omniscript RT kit, which we use to generate cDNA via reverse transcription for RT-qPCR.

24. While we have found that pooling cells from two starting wells to seed one new well of EBs works reliably, we recommend experimenting with different seeding densities (e.g., 3:2–3:1) to account for variations in initial iPSC confluency and cell line viability. Hence, we recommend plating out at least two wells of EBs, especially when attempting this assay for the first time.

25. The optimal EB seeding density for attachment may vary depending on the average size, shape, as well as total number of (viable) EBs. As a general guide, we plate EBs onto the same surface area the iPSC were harvested from (i.e., if two wells of iPSC were used to form 1 well of EBs, plate 1 well of EBs back onto 2 wells).

26. Cultures may achieve ~100% confluency within 14 days and still appear healthy when sufficient medium is provided. If substantial cell death is noted with "overcrowding" of germ layer cells despite sufficient feeding, we recommend repeating the assay using a lower EB seeding density on DIV4.

27. We typically use TaqMan probes in combination with the TaqMan Gene Expression Master Mix for routine RT-qPCR, adhering to manufacturer instructions for general preparation of reaction mixes and overall plate set-up.

28. Instead of adopting the standard curve method, we prefer a comparative C_T analysis. With every new assay run, we include three reference genes, i.e., *HPRT1*, *EEF2*, and *RPLP0*, which were selected based on the expression stability determined using RefFinder [47]. While we have found that *HPRT1* generally exhibited the most stable expression during differentiation of iPSC to embryoid bodies, we nevertheless recommend performing a preliminary assessment for a range of reference genes.

29. For the TaqMan pluripotency scorecard, RNA samples must be obtained from a minimum of 2 time points (i.e., DIV0 and DIV14). The abundance of EBs should be scaled up if additional time points are desired (e.g., DIV4, DIV7, DIV21).

30. STR analysis can be performed in-house by laboratories with suitable infrastructure or can be performed by third-party providers. We have previously used the Australian Genome Research Facility (Melbourne, Australia) to perform STR analysis on donor-matched CLN3 fibroblasts and iPSCs [1] and have also used this to confirm a 100% match between the CLN2 fibroblasts and iPSCs presented herein.

31. DNA purification can be performed directly using the PCR mixture if there is very minimal non-specific banding. Otherwise, we recommend manually isolating the target amplicon (e.g., cutting the desired band using a clean scalpel) and extracting the DNA from the excised gel fragment. For DNA purification and gel extraction, we prefer using the Monarch PCR & DNA Cleanup Kit as well as the Monarch DNA Gel Extraction Kit.

Acknowledgments

We are grateful to our participant and their family for providing a skin biopsy. We thank Helena Liang (Centre for Eye Research Australia) for technical assistance.

Funding for this work was provided by Batten Disease Support and Research Association (Australia) and the Medical Research Future Fund (MRF2023144). The University of Tasmania provided core laboratory facilities.

References

1. Chear S, Perry S, Wilson R et al (2022) Lysosomal alterations and decreased electrophysiological activity in CLN3 disease patient-derived cortical neurons. Dis Model Mech 15(12): dmm049651
2. Gomez-Giro G, Arias-Fuenzalida J, Jarazo J et al (2019) Synapse alterations precede neuronal damage and storage pathology in a human cerebral organoid model of CLN3-juvenile neuronal ceroid lipofuscinosis. Acta Neuropathol Commun 7(1):222
3. Tang C, Han J, Dalvi S et al (2021) A human model of Batten disease shows role of CLN3 in phagocytosis at the photoreceptor-RPE interface. Commun Biol 4(1):161
4. Basak I, Hansen RA, Ward ME et al (2021) Deficiency of the lysosomal protein CLN5 alters lysosomal function and movement. Biomolecules 11(10):1412
5. Lojewski X, Staropoli JF, Biswas-Legrand S et al (2014) Human iPSC models of neuronal ceroid lipofuscinosis capture distinct effects of TPP1 and CLN3 mutations on the endocytic pathway. Hum Mol Genet 23(8):2005–2022
6. Sato Y, Kobayashi H, Higuchi T et al (2017) Metabolomic profiling of Pompe disease-induced pluripotent stem cell-derived Cardiomyocytes reveals that oxidative stress is associated with cardiac and skeletal muscle pathology. Stem Cells Transl Med 6(1):31–39
7. Ter Huurne M, Parker BL, Liu NQ et al (2023) GLA-modified RNA treatment lowers GB3 levels in iPSC-derived cardiomyocytes from Fabry-affected individuals. Am J Hum Genet 110(9):1600–1605
8. Pham V, Sertori Finoti L, Cassidy MM et al (2024) A novel iPSC model reveals selective vulnerability of neurons in multiple sulfatase deficiency. Mol Genet Metab 141(2):108116
9. Frati G, Luciani M, Meneghini V et al (2018) Human iPSC-based models highlight defective glial and neuronal differentiation from neural progenitor cells in metachromatic leukodystrophy. Cell Death Dis 9(6):698
10. Hossain MA, Hasegawa-Ogawa M, Manome Y et al (2022) Generation and characterization of motor neuron progenitors and motor neurons using metachromatic leukodystrophy-induced pluripotent stem cells. Mol Genet Metab Rep 31:100852
11. Rabenstein M, Peter F, Joost S et al (2017) Decreased calcium flux in Niemann-Pick type C1 patient-specific iPSC-derived neurons due to higher amount of calcium-impermeable AMPA receptors. Mol Cell Neurosci 83:27–36
12. Pluvinage JV, Sun J, Claes C et al (2021) The CD22-IGF2R interaction is a therapeutic target for microglial lysosome dysfunction in Niemann-Pick type C. Sci Transl Med 13(622):eabg2919

13. Bayo-Puxan N, Terrasso AP, Creyssels S et al (2018) Lysosomal and network alterations in human mucopolysaccharidosis type VII iPSC-derived neurons. Sci Rep 8(1):16644
14. Hong J, Cheng YS, Yang S et al (2022) iPS-derived neural stem cells for disease modeling and evaluation of therapeutics for mucopolysaccharidosis type II. Exp Cell Res 412(1): 113007
15. Huang W, Cheng YS, Yang S et al (2021) Disease modeling for Mucopolysaccharidosis type IIIB using patient derived induced pluripotent stem cells. Exp Cell Res 407(1):112785
16. Canals I, Soriano J, Orlandi JG et al (2015) Activity and high-order effective connectivity alterations in Sanfilippo C patient-specific neuronal networks. Stem Cell Rep 5(4):546–557
17. Matsushita K, Numakawa T, Odaka H et al (2019) Presynaptic dysfunction in neurons derived from Tay-Sachs iPSCs. Neuroscience 414:128–140
18. Allende ML, Cook EK, Larman BC et al (2018) Cerebral organoids derived from Sandhoff disease-induced pluripotent stem cells exhibit impaired neurodifferentiation. J Lipid Res 59(3):550–563
19. Mangiameli E, Cecchele A, Morena F et al (2021) Human iPSC-based neurodevelopmental models of globoid cell leukodystrophy uncover patient- and cell type-specific disease phenotypes. Stem Cell Rep 16(6):1478–1495
20. Aguisanda F, Yeh CD, Chen CZ et al (2017) Neural stem cells for disease modeling of Wolman disease and evaluation of therapeutics. Orphanet J Rare Dis 12(1):120
21. Nakamura R, Tohnai G, Nakatochi M et al (2023) Genetic factors affecting survival in Japanese patients with sporadic amyotrophic lateral sclerosis: a genome-wide association study and verification in iPSC-derived motor neurons from patients. J Neurol Neurosurg Psychiatry 94(10):816–824
22. Czuppa M, Dhingra A, Zhou Q et al (2022) Drug screen in iPSC-Neurons identifies nucleoside analogs as inhibitors of (G(4)C(2))(n) expression in C9orf72 ALS/FTD. Cell Rep 39(10):110913
23. Fujimori K, Ishikawa M, Otomo A et al (2018) Modeling sporadic ALS in iPSC-derived motor neurons identifies a potential therapeutic agent. Nat Med 24(10):1579–1589
24. Rothstein JD, Baskerville V, Rapuri S et al (2023) G(2)C(4) targeting antisense oligonucleotides potently mitigate TDP-43 dysfunction in human C9orf72 ALS/FTD induced pluripotent stem cell derived neurons. Acta Neuropathol 147(1):1
25. Griffin TA, Anderson HC, Wolfe JH (2015) Ex vivo gene therapy using patient iPSC-derived NSCs reverses pathology in the brain of a homologous mouse model. Stem Cell Rep 4(5):835–846
26. Clarke D, Pearse Y, Kan SH et al (2018) Genetically corrected iPSC-derived neural stem cell grafts deliver enzyme replacement to affect CNS disease in Sanfilippo B mice. Mol Ther Methods Clin Dev 10:113–127
27. Doerr J, Bockenhoff A, Ewald B et al (2015) Arylsulfatase a overexpressing human iPSC-derived neural cells reduce CNS Sulfatide storage in a mouse model of metachromatic Leukodystrophy. Mol Ther 23(9):1519–1531
28. Morimoto S, Takahashi S, Ito D et al (2023) Phase 1/2a clinical trial in ALS with ropinirole, a drug candidate identified by iPSC drug discovery. Cell Stem Cell 30(6):766–780 e769
29. Warren L, Ni Y, Wang J et al (2012) Feeder-free derivation of human induced pluripotent stem cells with messenger RNA. Sci Rep 2:657
30. Warren L, Manos PD, Ahfeldt T et al (2010) Highly efficient reprogramming to pluripotency and directed differentiation of human cells with synthetic modified mRNA. Cell Stem Cell 7(5):618–630
31. Okita K, Matsumura Y, Sato Y et al (2011) A more efficient method to generate integration-free human iPS cells. Nat Methods 8(5): 409–412
32. Hong H, Takahashi K, Ichisaka T et al (2009) Suppression of induced pluripotent stem cell generation by the p53-p21 pathway. Nature 460(7259):1132–1135
33. Chou BK, Mali P, Huang X et al (2011) Efficient human iPS cell derivation by a non-integrating plasmid from blood cells with unique epigenetic and gene expression signatures. Cell Res 21(3):518–529
34. Drozd AM, Walczak MP, Piaskowski S et al (2015) Generation of human iPSCs from cells of fibroblastic and epithelial origin by means of the oriP/EBNA-1 episomal reprogramming system. Stem Cell Res Ther 6(1):122
35. Yu J, Hu K, Smuga-Otto K et al (2009) Human induced pluripotent stem cells free of vector and transgene sequences. Science 324(5928):797–801
36. Schlaeger TM, Daheron L, Brickler TR et al (2015) A comparison of non-integrating reprogramming methods. Nat Biotechnol 33(1):58–63
37. Ma L, Prada AM, Schmidt M et al (2021) Generation of pathogenic TPP1 mutations in human stem cells as a model for neuronal

ceroid lipofuscinosis type 2 disease. Stem Cell Res 53:102323

38. Gardner E, Mole SE (2021) The genetic basis of phenotypic heterogeneity in the neuronal ceroid Lipofuscinoses. Front Neurol 12: 754045

39. Talbot J, Chear S, Phipps A et al (2022) Image-based quantitation of Kainic acid-induced Excitotoxicity as a model of neurodegeneration in human iPSC-derived neurons. Methods Mol Biol 2549:187–207

40. King NE, Courtney JM, Brown LS et al (2024) Induced pluripotent stem cell derived pericytes respond to mediators of proliferation and contractility. Stem Cell Res Ther 15(1):59

41. Wang Q, Chear S, Wing K et al (2021) Use of CRISPR/Cas ribonucleoproteins for high throughput gene editing of induced pluripotent stem cells. Methods 194:18–29

42. Neavin D, Nguyen Q, Daniszewski MS et al (2021) Single cell eQTL analysis identifies cell type-specific genetic control of gene expression in fibroblasts and reprogrammed induced pluripotent stem cells. Genome Biol 22(1):76

43. Mehta A, Lu P, Taylor BV et al (2021) Generation of MNZTASi001-A, a human pluripotent stem cell line from a person with primary progressive multiple sclerosis. Stem Cell Res 57: 102568

44. Lin CF, Naj AC, Wang LS (2013) Analyzing copy number variation using SNP array data: protocols for calling CNV and association tests. Curr Protoc Hum Genet 79:1 27 21–21 27 15

45. Wang K, Li M, Hadley D et al (2007) PennCNV: an integrated hidden Markov model designed for high-resolution copy number variation detection in whole-genome SNP genotyping data. Genome Res 17(11): 1665–1674

46. Colella S, Yau C, Taylor JM et al (2007) QuantiSNP: an objective Bayes hidden-Markov model to detect and accurately map copy number variation using SNP genotyping data. Nucleic Acids Res 35(6):2013–2025

47. Xie F, Xiao P, Chen D et al (2012) miRDeepFinder: a miRNA analysis tool for deep sequencing of plant small RNAs. Plant Mol Biol 80:75

Chapter 13

Using CMAC Staining for Vacuole Characterization in Yeast

José Ángel Clemente-Ramos and Sara E. Mole

Abstract

CMAC (7-amino-4-chloromethylcoumarin) staining is a valuable tool for visualizing acidic organelles, including vacuoles, in yeast cells. By selectively accumulating in acidic compartments, CMAC allows specific labeling and visualization of vacuoles in living or fixed cells, aiding in the investigation of vacuole dynamics, morphology, and distribution under various conditions. This staining method provides a clear contrast against cellular autofluorescence, enhancing imaging clarity. It can be combined with automated imaging capture for faster throughput and analysis.

Characterizing vacuole phenotypes using CMAC staining contributes to our understanding of cellular homeostasis and disease mechanisms in yeast. It has applications in studying fundamental cellular processes, including endocytic trafficking and autophagy, as well as in drug screening assays, and can be used in yeast models of disease to offer insights into potential therapeutic targets for diseases affecting the lysosome, including neurodegenerative diseases.

Key words CMAC staining, 7-amino-4-chloromethylcoumarin, Fission yeast, *Schizosaccharomyces pombe*, Lysosome, Vacuole, Homeostasis

1 Introduction

Fluorescent staining techniques play a pivotal role in various fields of biomedical research, particularly in the investigation of cellular structures, biomolecules, and cellular processes. Among these techniques, CMAC (7-amino-4-chloromethylcoumarin) staining has emerged as a valuable tool for visualizing cellular compartments, particularly acidic organelles such as lysosomes and endosomes, and vacuoles, their equivalent structures, in yeast and plant cells. These compounds readily diffuse across cell membranes due to their hydrophobic nature. The chloromethyl group of CMAC reacts with thiols such as glutathione, producing fluorescent conjugates that are membrane-impermeable and so retained within the vacuole lumen, making it suitable for fluorescence microscopy, enzymatic and viability assays, and flow cytometry applications [1–4].

Stephanie M. Hughes and Indranil Basak (eds.), *Lysosomes: Methods and Protocols*, Methods in Molecular Biology, vol. 2976, https://doi.org/10.1007/978-1-0716-4844-5_13,

The use of CMAC staining is grounded in the principles of cell biology and fluorescence spectroscopy. CMAC can be used to visualize acidic compartments within cells due to its pH-dependent fluorescence properties. In acidic environments, CMAC undergoes protonation, generating a fluorescent derivative that is trapped in the vacuole lumen, resulting in increased fluorescence emission. This property allows for the specific labeling and visualization of acidic organelles within living or fixed cells in a range of fungi [5–11]. We use fission yeast *Schizosaccharomyces pombe* as a valuable single-cell model to give more insight into lysosomal diseases, especially the neuronal ceroid lipofuscinoses (NCL, Batten disease) [12–18].

CMAC staining offers several advantages over traditional staining methods. Its cell-permeable nature facilitates the staining of live cells without the need for cell permeabilization procedures, thereby preserving cellular integrity and physiological conditions. Additionally, the blue fluorescence emission of CMAC provides distinct contrast against cellular autofluorescence and background noise, enhancing the clarity and specificity of imaging.

In research and clinical settings, CMAC staining has been extensively utilized to study various cellular processes and pathologies associated with lysosomal dysfunction, endocytic trafficking, and autophagy. Furthermore, its compatibility with high-throughput screening techniques makes CMAC staining a valuable tool for drug discovery and therapeutic development targeting lysosomal disorders and related diseases [19, 20].

2 Materials

2.1 CMAC Staining Solution

These two commercially available compounds can be used interchangeably to carry out staining: Cell tracker™ Blue CMAC Dye (Molecular Probes/Invitrogen/ThermoFisher Scientific, C2110) and CellView Blue CMAC (ABP Biosciences, C037). Both reagents should be stored at −20 °C, kept desiccated, and protected from light. Guidelines for preparing the reagents:

1. The reagent should be allowed to warm to room temperature before opening the vial. Prepare a stock concentration of 10 mM by dissolving 5 mg of CMAC in 2.3852 mL of high-quality, anhydrous DMSO. Aliquot (100 μL) and store protected from light and moisture at −20 °C. When stored properly, both solids and resuspended stock solutions are stable for at least 6 months.
2. Avoid frequent freezing and thawing of reagents. The optimal working concentration of the probe for staining varies depending on the application. Testing a range of concentrations from 25 to 100 μM is recommended. In general, short-term

experiments using fission yeast or rapidly dividing cells requires a working concentration of 100 μM. To maintain normal cellular physiology and reduce potential artifacts, keep the dye concentration as low as possible.

2.2 Yeast Growth Media

S. pombe yeast cells were grown in a standard rich yeast growth medium (YES, Yeast Extract with Supplements: Yeast Extract, Dextrose, and Supplements of Adenine, Histidine, Leucine, Uracil, and Lysine) or Edinburgh Minimal Medium (EMM) containing appropriate supplements. Media, growth, maintenance of strains, and genetic methods are previously described [21–23]. Cells were incubated at the chosen temperature (wild-type strains are generally grown at 29–32 °C) using an orbital shaker incubator (e.g., New Brunswick Innova 4230) to allow cell aeration (225 rpm).

2.3 Incubation Buffer

After having concentrated the yeast cultures by centrifugation, the cells are resuspended in 10 mM Na-HEPES pH 7.4 buffer containing 5% glucose. This can be sterilized by filtration and stored at 4 °C until use.

2.4 Wash Buffer

PBS (phosphate-buffered saline): A buffer solution containing phosphate salts and sodium chloride, typically prepared at pH 7.4 and sterilized by autoclaving. Used for washing yeast cells before and after staining to remove excess dye and minimize background fluorescence.

2.5 Microscopy Equipment

Images were acquired using a Leica SPE2 true scanning confocal microscope with a 63×/1.4 Oil immersion objective, and recordings were made utilizing Leica AF software. Additionally, a Zeiss LSM 710 laser scanning confocal microscope equipped with a PlanApochromat 63×/1.40 Oil DIC M27 objective was utilized, and subsequent image processing was performed using ZEN 2009 software. Excitation of CMAC was achieved through a 405-nm UV diode/laser, with emission acquired within the 410–495 nm spectral window as the blue channel. Optical sections were captured at a thickness of 1 μm. For yeast cells, a 63×/100× oil-immersion objective is considered ideal.

Alternatively, a Zeiss Observer widefield fluorescence inverted microscope equipped with a Hamamatsu Flash 4.0v3 camera and a Plan Apo 63×/1.40 oil PH3 objective, along with appropriate filter sets for excitation and emission wavelengths for CMAC dye (typically around 360–380 nm for excitation and 420–470 nm for emission), was utilized. Images were processed for brightness, contrast, and/or sharpness adjustments using ImageJ/Fiji software [24].

3 Methods

Vacuole staining using CMAC is a widely used technique to visualize vacuole dynamics and morphology in yeast cells. This protocol outlines the steps required to carry out CMAC staining in fission yeast for fluorescence microscopy analysis. Adjustments to specific concentrations and conditions may be required based on experimental requirements.

3.1 Preparation of Yeast Culture

1. Inoculate a single colony or overnight patch of the selected *S. pombe* strain into 5 mL of liquid medium (YES or EMM complete with appropriate supplements) from fresh agar plates previously incubated for 1–2 days at the appropriate temperature until cells are in early stationary phase.
2. Incubate yeast culture overnight (10–12 h) at 30 °C in a shaking incubator. This preculture can be used to inoculate a larger culture (*see* **Note 1**).
3. Transfer 1 mL of the overnight culture into 10 mL of fresh YES or EMM liquid medium and incubate at 30 °C with shaking until reaching log-phase (OD600 = 0.8–1.0) ($\cong 1.0 \times 10^7$ cells/mL), taking into consideration the usual generation times for the strains being studied (*see* **Note 2**).

3.2 CMAC Staining of Yeast Cells

1. Harvest 1 mL of exponential phase growing cells by centrifugation (900× *g*) for 1 min at room temperature in a benchtop centrifuge. Discard the supernatant (*see* **Note 3**).
2. Wash the cell pellet with 1 mL of 10 mM HEPES buffer pH 7.4 containing 5% glucose (*see* **Note 4**).
3. Resuspend the cell pellet (1.0×10^7 cells/mL) in 1 mL of 10 mM HEPES buffer pH 7.4 containing 5% glucose.
4. Add CMAC dye to the cell suspension to a final concentration of 100 μM, using the 10 mM DMSO stock solution (e.g., 10 μL of CellTracker™ Blue CMAC to 1 mL of the cell suspension) (*see* **Note 5**).
5. Incubate the cell suspension with CMAC dye in the dark for 15–30 min at 30 °C without agitation (*see* **Note 6**).
6. Wash cells once with 1 mL of 1× PBS to remove unbound dye.
7. Finally, collect the cells by centrifugation (1 min at 900× *g*), resuspend in a small amount of growth medium, and visualize immediately using confocal microscopy.

3.3 Microscopy and Short-Term Imaging

1. Pipet an aliquot (1.7–2.0 μL) of the stained fission yeast cells onto the microscope slide.
2. Cover the sample with a coverslip (*see* **Note 7**).

3. Visualize the stained yeast cells using a confocal microscope equipped with appropriate filter sets for CMAC dye (excitation: 353–354 nm, emission: 463–469 nm) using a 63×/100× oil-immersion objective.
4. Capture fluorescence images corresponding to the middle planes of the stained yeast cells using suitable imaging software (*see* **Note 8**).

3.4 Image Preprocessing, Segmentation, and Measurement of Vacuoles

1. Use image processing software (e.g., ImageJ/Fiji) to preprocess the acquired images.
2. Apply appropriate filters (e.g., Gaussian blur) to reduce noise and enhance image quality.
3. Adjust brightness and contrast settings to optimize visualization of vacuoles while minimizing background signal.
4. Utilize automated segmentation algorithms or manual thresholding techniques to identify and outline individual vacuoles in the images (*see* **Note 9**) (Fig. 1).

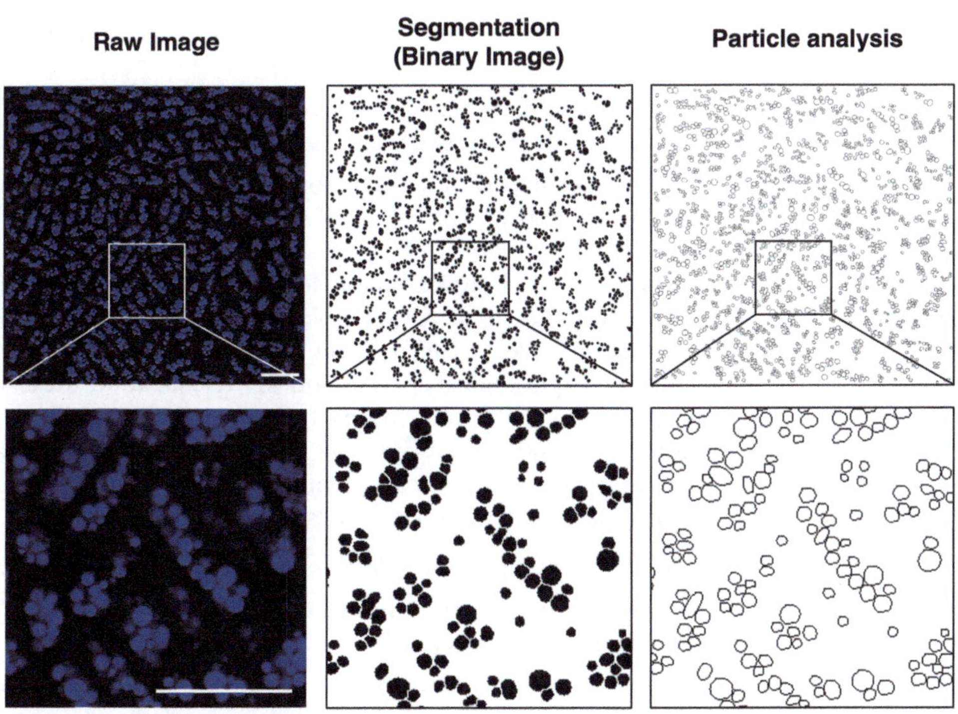

Fig. 1 Pipeline for *S. pombe* vacuole segmentation using the manual thresholding technique. Cells were grown in minimal media (EMM) overnight to log phase and stained with CellTracker Blue CMAC to visualize the vacuole lumen. Particles were identified and analyzed to calculate their area using calibrated images. Bar, 10 μm

3.5 Manual Thresholding Technique

1. Open the acquired images in ImageJ/Fiji.
2. Set the scale according to the resolution of the images obtained during the acquisitions and the objective used (63× or 100×) to determine the px/microns ratio (*Analyze* > *Set Scale*) (*see* **Note 10**).
3. Convert the images to grayscale (*Image* > *Lookup Tables* > *Grays*).
4. Apply local thresholding using the Bernstein method with a radius set to 5–15 (*Image* > *Adjust* > *AutoLocal Threshold*).
5. Convert the image to binary (*Process* > *Binary* > *Make Binary*).
6. Invert the LUT (*Image* > *Lookup Tables* > *Invert LUT*) when necessary. In the final binary image for quantification, the vacuoles should have values equal to 1 (255 in the histogram) and the background a value equal to zero.
7. Use the Wand (tracing) tool to manually clean unwanted points (*Edit* > *Options* > *Wand Tool*).
8. Apply watershed to separate closely located particles (*Process* > *Binary* > *Watershed*).
9. If necessary, divide merged vacuoles into individual vacuoles manually (*see* **Note 11**).
10. Analyze particles under (*Analyze* > *Analyze Particles*) (*see* **Note 12**).
11. Export the analyzed data to your spreadsheet software (e.g., Excel) for further analysis.
12. Calculate the diameter (d) from the area (A) of the particles using the formula: $d = 2 \times \sqrt{(A/\pi)}$.

3.6 Automatic Segmentation Methods

There are several plugins available for ImageJ/Fiji that offer a wide range of functionalities for vacuole quantification in fission yeast stained with CMAC, from basic segmentation and measurement to advanced morphological analysis and 3D visualization. Researchers can select and combine plugins based on their specific analysis requirements and preferences (*see* **Note 13**). Using the StarDist Plugin in Image J/Fiji, the segmented objects need to be processed to obtain particle descriptors as described below.

1. Start the plugin from (*Plugins* > *StarDist* > *StarDist 2D*).
2. Select a neural network model from the dropdown list: "Versatile (fluorescent nuclei)."
3. Adjust the NMS (Non-Maximum Suppression) postprocessing parameters depending on the reliability of detecting particles in the sample as well as their degree of overlap. If in doubt, load the default NMS parameters of the selected built-in model.

4. The output with segmented objects can be returned as a Label Image or in the ROI Manager and needs to be processed to obtain particle descriptors as previously described.
5. Convert the image to 8-bits (*Image* > *Type* > *8-bits*).
6. Select all the particles in the label image, setting the threshold to 1–255 (*Image* > *Adjust* >*Threshold*) and select Apply to convert the image to binary.
7. Apply watershed to separate closely located particles (*Process* > *Binary* > *Watershed*).
8. Analyze particles under (*Analyze* > *Analyze Particles*) (*see* **Note 12**).
9. Export the analyzed data to your spreadsheet software (e.g., Excel) for further analysis.
10. Calculate the diameter (d) from the area (A) of the particles using the formula: $d = 2 \times \sqrt{(A/\pi)}$.

3.7 Statistical Analysis and Visualization

1. Perform statistical analysis to compare vacuole morphology between different genetic perturbations or experimental conditions. Use appropriate statistical tests. In instances where only two samples were compared, significance was determined using an unpaired t-test. In instances where more than two strains were compared, assuming a normal distribution, an ordinary one-way ANOVA was used with a Tukey's test to determine significant differences in vacuole size and shape parameters. Kruskal–Wallis test for ranks was performed as a nonparametric alternative to ANOVA, accompanied by Dunn's multiple comparison test to calculate P value between samples. A minimum of three replicates for each strain/condition were analyzed, imaging at least 500 cell lumen areas in each replica.
2. Generate visual representations of vacuole morphology data, such as scatter plots, histograms, or preferably through SuperPlots (Fig. 2), communicating both the cell-level variability and the experimental reproducibility [25]. Present quantitative results alongside representative images to illustrate vacuole staining and morphology changes effectively (Fig. 3).

4 Notes

1. Cell growth can be monitored by measuring the A_{600} of early log-phase cell populations using an Eppendorf BioPhotometer Model #6131 spectrophotometer (Eppendorf AG, Hamburg; A_{600} 0.1 = 1 × 10^6 cells/mL). Using disposable inoculation loops (5 μL) to take a scoop of cells for inoculum, *S. pombe* cultures will reach growth in the late logarithmic or stationary phase after 12 h at 30 °C, while shaking at 225 rpm,

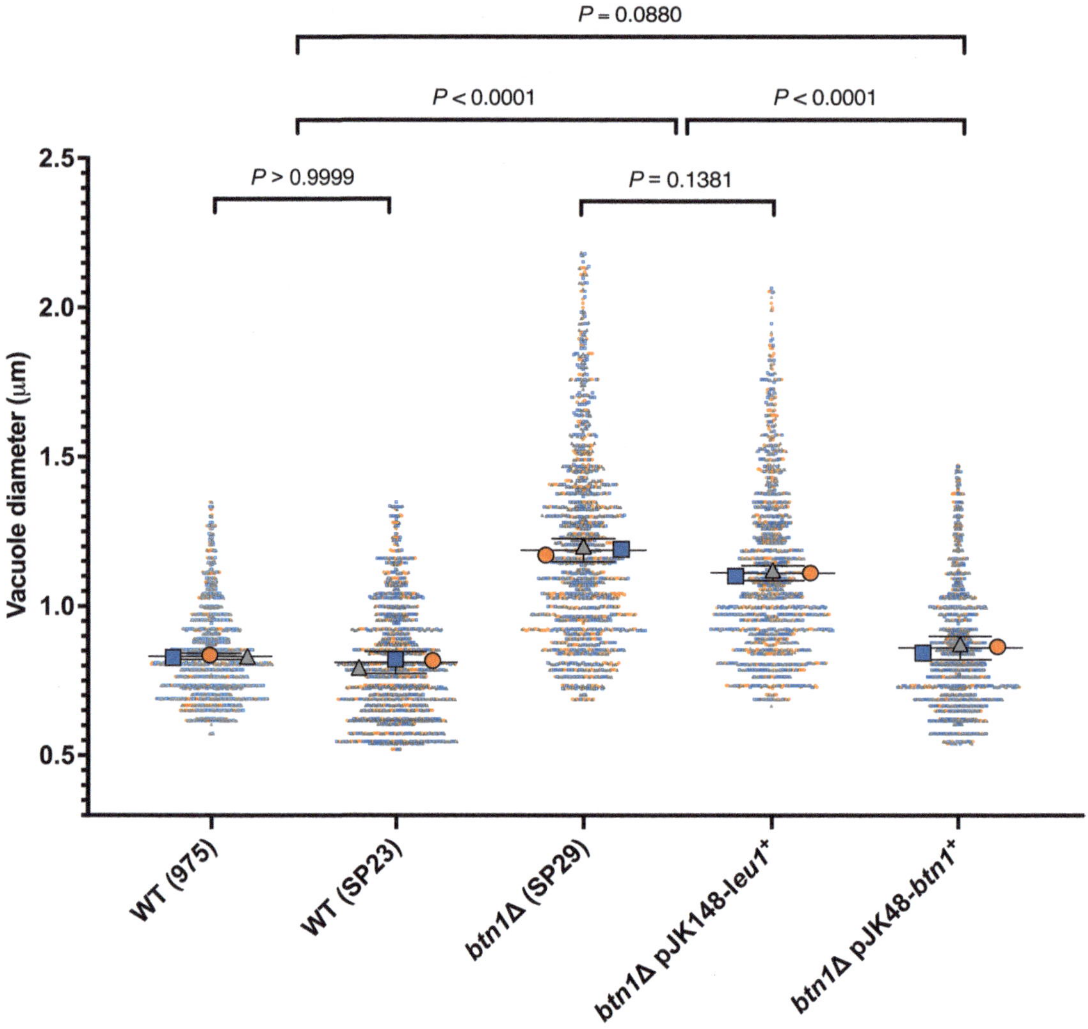

Fig. 2 Quantification of the vacuole diameter measured in the indicated strains. Note that deletion of the endogenous gene *btn1*$^+$ in wild-type strain SP23 causes a significant increase in vacuole size and ectopic expression of *btn1*$^+$ in cells lacking *btn1*$^+$ causes a significant reduction in vacuole size. Data is representative of three experiments, for each vacuoles measured n $>$ 500. The mean, standard deviation, and statistical significance, determined by Dunn's multiple comparisons tests after analysis of variance for ranks (Kruskal-Wallis Test), are shown for each pairwise comparison. $P >$ 0.05 (non-significant), $P \leq$ 0.0001 (****)

yielding OD600 nm values of around 2–3. For physiological experiments, it is important that cultures are maintained in mid-early log-phase growth between 5×10^6 and 1×10^7 cells/mL.

2. Depending on the culture medium, the temperature, and the physiological state or adaptation to the medium, the generation time for *S. pombe* can range between 2 and 3 h. An

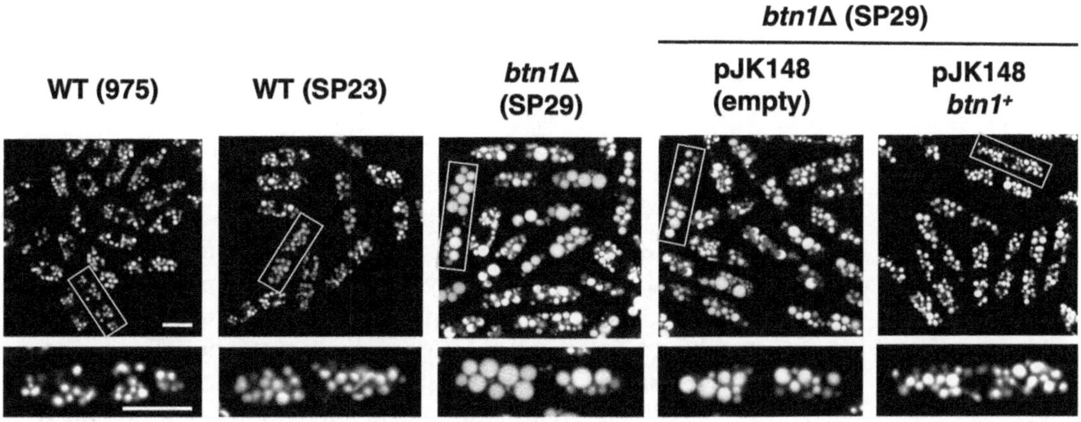

Fig. 3 Confocal fluorescence microscopy images of live *S. pombe* cells growing in EMM at 30°C with CellTracker Blue CMAC stained vacuoles (100 μM). Strains shown are the wild-type reference strain (WT 975), wild-type control strain (WT SP23), strain *btn1*Δ (SP29), strain *btn1*Δ containing empty integrative plasmid pJK148-*leu1*$^+$, and strain *btn1*Δ containing pJK148-*btn1*$^+$ that restores the expression of *btn1*$^+$ from its endogenous promoter at *leu1*$^+$ loci. Bar, 5 μm

adaptation period of no less than 1 h must be considered to allow the cells to recover from the stationary phase.

3. The centrifugation conditions should always be the gentlest that allow the cells to be pelleted. Keep these below 3–5 min.
4. If working in YES, wash the cells twice with EMM complete to reduce the background before continuing with the protocol. Alternatively, EMM can be used as a washing buffer before proceeding with the next steps of the staining protocol.
5. The duration of incubation serves as a guideline and depends on the cell population density. A staining duration of just 10 min yields a uniformly stained sample of 1 mL from *S. pombe* cultures in logarithmic phase. The washing step is key to reducing the excess of dye within the medium and diluting any autofluorescence present in the cytosol. While CMAC exhibits accumulation within vacuoles and fluoresces in response to the pH environment, it may permeate into the cytoplasm when the integrity of the vacuole membrane is compromised. Adjust the concentration of CMAC dye and the duration of staining to achieve an optimal signal-to-background ratio.
6. Alternatively, after washing with 10 mM HEPES buffer pH 7.4 containing 5% glucose, the cell suspension can be incubated for 30 min in 500 μL of EMM with a final dye concentration of 100 μM (added from a 10 mM solution in DMSO) on a vertical orbital shaker at room temperature and in the dark. The cells can then be centrifuged (900× *g*) for 1 min at room temperature, resuspended in 1 mL of EMM, and incubated for 10 min

under the previous conditions. Finally, collect the cells by centrifugation (1 min at 900× *g*) and resuspend in a small amount of medium to proceed with visualization under the microscope [26, 27].

7. The use of any nail polish-type compound to seal the edges to prevent drying is not recommended. A cell volume of 1.8 μL allows *S. pombe* cells to be distributed in a monolayer when using 18 × 18 mm coverslips. To avoid the formation of bubbles and achieve a monolayer of cells on the microscope slide, it may be necessary to clean it with 70% ethanol and wipe dry with a tissue to remove dust and residue. The cells in a monolayer can be kept up to 5 min in physiological conditions, especially when cells are suspended in the culture medium (EMM complete). After that time, artifacts produced by cellular stress become evident on the microscope slide, so a fresh slide from the concentrated cell suspension needs to be prepared.
8. Ensure consistent imaging parameters such as exposure time, laser intensity, gain setting, magnification, and resolution across all images.
9. Apply suitable segmentation methods based on vacuole morphology, such as intensity-based thresholding or edge detection algorithms. Validate the accuracy of vacuole segmentation by comparing segmented regions with original images and adjusting parameters as needed.
10. Depending on the software and imaging platforms, this information may be recorded in the image metadata. When the scale is adjusted properly, keep the checkbox "Global" clicked (*Analyze > Set Scale*) so that the changes affect all images open in your session.
11. To manually perform the separation of vacuoles that are in contact, use the Line selection to create a straight line that divides the particles, then draw the line (*Edit > Draw*). Alternatively, use Drawing Tools (pencil) to identify isolated particles that the watershed option cannot resolve. Set the drawing width value to 1 pixel. Set the foreground and background colors by double-clicking on the eye dropper (color picker) (*Image > Color*). Choose the same color as the background in the binary image.
12. Ensure "Area," "Feret's Diameter," "Perimeter," and "Shape Descriptors" are selected under (*Analyze > Set Measurements*). Adjust the decimal place to 4. Feret's diameter offers a quick estimate of particle size but may overestimate for irregular shapes, while area measurements provide more accurate and comprehensive size information but are computationally intensive and sensitive to image quality. Optionally, filter particle size

to exclude small/large staining artifacts (e.g., set Size range from 0.10 to 2 μm^2 and Circularity to 0.75) under (*Analyze* > *Analyze Particles*). The "Display results," "Add to Manager," and "Exclude on edges" checkboxes need to be selected.

13. Here are some commonly used plugins:
    - MorphoLibJ (https://imagej.net/plugins/morpholibj) [28] is a collection of mathematical morphology methods and plugins for image analysis in ImageJ/Fiji under the *Plugins › MorphoLibJ › Binary Images.*
    - 3D ImageJ Suite (https://imagej.net/plugins/3d-imagej-suite/) [29] with plugins such as "3D Object Counter" and "3D Viewer" enable volumetric analysis and visualization of vacuoles in 3D space.
    - StarDist (https://imagej.net/plugins/stardist) [30] is an automated detection and segmentation method for identifying cells and nuclei using star-convex polygons. The plugin offers the capability to deploy pre-trained models on novel images, particularly those featuring vacuole/nucleus structures. This feature is especially advantageous for images characterized by challenging conditions, such as densely packed cells with vacuole particles, where the typically spherical shapes are prevalent.

Final Note

The major sources of issues during CMAC staining in yeast and strategies to identify and overcome them:

1. Insufficient staining: If yeast cells are not adequately stained with CMAC, it may be challenging to visualize vacuoles. Overcoming: Increase the concentration of CMAC dye used for staining or prolong the staining incubation time. Consider performing an alternative staining protocol with an incubation period of 30 min (as described in **Note 6**), then washing and using an additional incubation for 10 min in culture medium before concentrating the cells again by centrifugation for visualization. Ensure that the CMAC stock solution is freshly prepared and stored appropriately to maintain its efficacy.
2. High background fluorescence: Excessive background fluorescence can obscure the signal from stained vacuoles, reducing image clarity and making it difficult to distinguish specific subcellular structures. Overcoming: Thoroughly wash the stained yeast cells with PBS after staining to remove excess dye. Additionally, adjust the concentration of CMAC dye and the duration of staining to achieve an optimal signal-to-background ratio.

3. Photobleaching: Prolonged exposure to excitation light can lead to photobleaching of CMAC dye, resulting in reduced fluorescence intensity over time. Overcoming: Minimize exposure time to excitation light by using appropriate filter sets and imaging settings. Utilize anti-fade mounting media containing antioxidants or oxygen scavengers to slow down photobleaching. Capture images using the lowest possible excitation intensity and optimize imaging parameters to minimize photobleaching while maintaining signal intensity. After 5 min, use a freshly prepared slide of cells.
4. Extreme cell conditions: When the pH of the vacuoles in yeast is not acidic or when the permeability of the vacuole membrane is altered, the staining pattern using CMAC may be affected. Different artifacts may be detected, such as a reduced or altered staining pattern affecting the accuracy and reliability of vacuole visualization [31, 32]. Overcoming: Validate staining protocols using wild-type cells grown under appropriate conditions to ensure competency. Then consider factors unique to the experimental strains and conditions to enable accurate interpretation of experimental results.

Acknowledgments

This work was supported by the National Institute for Health Research Biomedical Research Centre at University College London, Great Ormond Street Institute of Child Health, and the USA Children's Brain Diseases Foundation. The authors thank previous laboratory members for their contributions in developing and testing the protocol (Elisa Tinelli, Yaxuan Lyu). S.E.M. receives salary support from University College London (UCL).

References

1. Niapour M, Berger S (2007) Flow cytometric measurement of calpain activity in living cells. Cytometry A 71A(7):475–485. https://doi.org/10.1002/cyto.a.20399
2. Johnson S, Nguyen V, Coder D (2013) Assessment of cell viability. Curr Protoc Cytom 64(1):9.2.1–9.2.26. https://doi.org/10.1002/0471142956.cy0902s64
3. Haugland RP, Spence MTZ, Johnson ID (1996) Handbook of fluorescent probes and research chemicals, 6th edn. Molecular Probes
4. Lawrence RE, Zoncu R (2019) The lysosome as a cellular Centre for signalling, metabolism and quality control. Nat Cell Biol 21(2): 133–142. https://doi.org/10.1038/s41556-018-0244-7
5. Stefan CJ, Blumer KJ (1999) A Syntaxin homolog encoded by VAM3 mediates Down-regulation of a yeast G protein-coupled receptor. J Biol Chem 274(3):1835–1841. https://doi.org/10.1074/jbc.274.3.1835
6. Cole L, Orlovich DA, Ashford AE (1998) Structure, function, and motility of vacuoles in filamentous fungi. Fungal Genet Biol 24(1–2):86–100. https://doi.org/10.1006/fgbi.1998.1051
7. Ohneda M, Arioka M, Nakajima H, Kitamoto K (2002) Visualization of vacuoles in Aspergillus oryzae by expression of CPY-EGFP. Fungal

Genet Biol 37(1):29–38. https://doi.org/10.1016/s1087-1845(02)00033-6

8. Shoji J-Y, Arioka M, Kitamoto K (2006) Vacuolar membrane dynamics in the filamentous fungus aspergillus oryzae. Eukaryot Cell 5(2):411–421. https://doi.org/10.1128/ec.5.2.411-421.2006
9. Haugland RP (1995) Detecting enzymatic activity in cells using Fluorogenic substrates. Biotech Histochem 70(5):243–251. https://doi.org/10.3109/10520299509108201
10. Cole L, Hyde GJ, Ashford AE (1997) Uptake and compartmentalisation of fluorescent probes byPisolithus tinctorius hyphae: evidence for an anion transport mechanism at the tonoplast but not for fluid-phase endocytosis. Protoplasma 199(1–2):18–29. https://doi.org/10.1007/bf02539802
11. Stewart A, Deacon JW (1995) Vital fluorochromes as tracers for fungal growth studies. Biotech Histochem 70(2):57–65. https://doi.org/10.3109/10520299509108318
12. Codlin S, Haines RL, Burden JJ, Mole SE (2008) Btn1 affects cytokinesis and cell-wall deposition by independent mechanisms, one of which is linked to dysregulation of vacuole pH. J Cell Sci 121(Pt 17):2860–2870. https://doi.org/10.1242/jcs.030122
13. Codlin S, Haines RL, Mole SE (2008) btn1 affects endocytosis, polarization of sterol-rich membrane domains and polarized growth in Schizosaccharomyces pombe. Traffic 9(6):936–950. https://doi.org/10.1111/j.1600-0854.2008.00735.x
14. Codlin S, Mole SE (2009) S. pombe btn1, the orthologue of the Batten disease gene CLN3, is required for vacuole protein sorting of Cpy1p and Golgi exit of Vps10p. J Cell Sci 122(Pt 8):1163–1173. https://doi.org/10.1242/jcs.038323
15. Gachet Y, Codlin S, Hyams JS, Mole SE (2005) btn1, the Schizosaccharomyces pombe homologue of the human Batten disease gene CLN3, regulates vacuole homeostasis. J Cell Sci 118(Pt 23):5525–5536. https://doi.org/10.1242/jcs.02656
16. Haines RL, Codlin S, Mole SE (2009) The fission yeast model for the lysosomal storage disorder Batten disease predicts disease severity caused by mutations in CLN3. Dis Model Mech 2(1–2):84–92. https://doi.org/10.1242/dmm.000851
17. Kitzmuller C, Haines RL, Codlin S, Cutler DF, Mole SE (2008) A function retained by the common mutant CLN3 protein is responsible for the late onset of juvenile neuronal ceroid lipofuscinosis. Hum Mol Genet 17(2):303–312. https://doi.org/10.1093/hmg/ddm306
18. Phillips SN, Muzaffar N, Codlin S, Korey CA, Taschner PE, de Voer G, Mole SE, Pearce DA (2006) Characterizing pathogenic processes in batten disease: use of small eukaryotic model systems. Biochim Biophys Acta 1762(10):906–919. https://doi.org/10.1016/j.bbadis.2006.08.010
19. Tournu H, Carroll J, Latimer B, Dragoi A-M, Dykes S, Cardelli J, Peters TL, Eberle KE, Palmer GE (2017) Identification of small molecules that disrupt vacuolar function in the pathogen Candida albicans. PLoS One 12(2):e0171145. https://doi.org/10.1371/journal.pone.0171145
20. Stolp ZD, Kulkarni M, Liu Y, Zhu C, Jalisi A, Lin S, Casadevall A, Cunningham KW, Pineda FJ, Teng X, Hardwick JM (2022) Yeast cell death pathway requiring AP-3 vesicle trafficking leads to vacuole/lysosome membrane permeabilization. Cell Rep 39(2):110647. https://doi.org/10.1016/j.celrep.2022.110647
21. Forsburg SL, Rhind N (2006) Basic methods for fission yeast. Yeast 23(3):173–183. https://doi.org/10.1002/yea.1347
22. Alfa CE, Gallagher IM, Hyams JS (1993) Antigen localization in fission yeast. Methods Cell Biol 37:201–222. https://doi.org/10.1016/s0091-679x(08)60251-4
23. Moreno S, Klar A, Nurse P (1991) Molecular genetic analysis of fission yeast *Schizosaccharomyces pombe*. Methods Enzymol 194:795–823. PMID: 2005825
24. Schindelin J, Arganda-Carreras I, Frise E, Kaynig V, Longair M, Pietzsch T, Preibisch S, Rueden C, Saalfeld S, Schmid B, Tinevez J-Y, White DJ, Hartenstein V, Eliceiri K, Tomancak P, Cardona A (2012) Fiji: an open-source platform for biological-image analysis. Nat Methods 9(7):676–682. https://doi.org/10.1038/nmeth.2019
25. Lord SJ, Velle KB, Mullins RD, Fritz-Laylin LK (2020) SuperPlots: communicating reproducibility and variability in cell biology. J Cell Biol 219(6). https://doi.org/10.1083/jcb.202001064
26. Yanguas F, Moscoso-Romero E, Valdivieso M-H (2019) Ent3 and GGA adaptors facilitate diverse anterograde and retrograde trafficking events to and from the prevacuolar endosome. Sci Rep 9(1). https://doi.org/10.1038/s41598-019-47035-5
27. Davidson R, Pontasch JA, Wu J-Q (2016) Sbg1 is a novel regulator for the localization

of the β-glucan synthase Bgs1 in fission yeast. PLoS One 11(11):e0167043

28. Legland D, Arganda-Carreras I, Andrey P (2016) MorphoLibJ: integrated library and plugins for mathematical morphology with ImageJ. Bioinformatics 32(22):3532–3534. https://doi.org/10.1093/bioinformatics/btw413
29. Ollion J, Cochennec J, Loll F, Escudé C, Boudier T (2013) TANGO: a generic tool for high-throughput 3D image analysis for studying nuclear organization. Bioinformatics 29(14): 1840–1841. https://doi.org/10.1093/bioinformatics/btt276
30. Schmidt U, Weigert M, Broaddus C, Myers G (2018) Cell detection with star-convex polygons. In: Medical image computing and computer assisted intervention—MICCAI 2018. Springer International Publishing, pp 265–273. https://doi.org/10.1007/978-3-030-00934-2_30
31. Karim MA, Mcnally EK, Samyn DR, Mattie S, Brett CL (2018) Rab-effector-kinase interplay modulates Intralumenal fragment formation during vacuole fusion. Dev Cell 47(1):80–97. e86. https://doi.org/10.1016/j.devcel.2018.09.002
32. Llinares E, Barry AO, André B (2015) The AP-3 adaptor complex mediates sorting of yeast and mammalian PQ-loop-family basic amino acid transporters to the vacuolar/lysosomal membrane. Sci Rep 5(1):16665. https://doi.org/10.1038/srep16665

Chapter 14

Modeling Lysosomal Disease in *Dictyostelium discoideum*: Examining the Trafficking and Secretion of Lysosomal Enzymes

William D. Kim and Robert J. Huber

Abstract

Non-mammalian models are powerful systems for enhancing our understanding of lysosomal function and lysosomal storage diseases. The social amoeba *Dictyostelium discoideum* is an excellent model organism for studying lysosomal function, as its genome encodes many proteins associated with lysosomal disease. Methods for gene knockout are straightforward in *D. discoideum* and include restriction enzyme-mediated integration (REMI) mutagenesis, homologous recombination via the Cre-loxP system, and CRISPR/Cas9-mediated gene editing, which collectively allow researchers to study protein function (e.g., lysosomal enzymes) in a genetically tractable biomedical model system. Additionally, activity assays for conserved lysosomal enzymes are well-established in *D. discoideum*. In this chapter, we outline methods for studying the intracellular localization and secretion of conserved lysosomal proteins in *D. discoideum*.

Key words *Dictyostelium discoideum*, Enzyme, Immunofluorescence, Lysosomes, Model organism, Neurodegeneration, Secretion, Trafficking, Western blotting

Abbreviations

CprG	cysteine protease G
CsaA	contact site A glycoprotein
ECL	enhanced chemiluminescence
ER	endoplasmic reticulum
IF	immunofluorescence
LoFlo	low fluorescence
M6P	mannose-6-phosphate
NCL	neuronal ceroid lipofuscinosis
REMI	restriction enzyme-mediated integration
VatC	V-ATPase subunit C
WGA	wheat germ agglutinin
WT	wild type

Stephanie M. Hughes and Indranil Basak (eds.), *Lysosomes: Methods and Protocols*, Methods in Molecular Biology, vol. 2976, https://doi.org/10.1007/978-1-0716-4844-5_14,

1 Introduction

1.1 Life Cycle and Culturing of D. discoideum

Dictyostelium discoideum is an amoeba that has a life cycle composed of distinct unicellular and multicellular phases (Fig. 1) [1]. During its unicellular phase, haploid amoebae feed on a

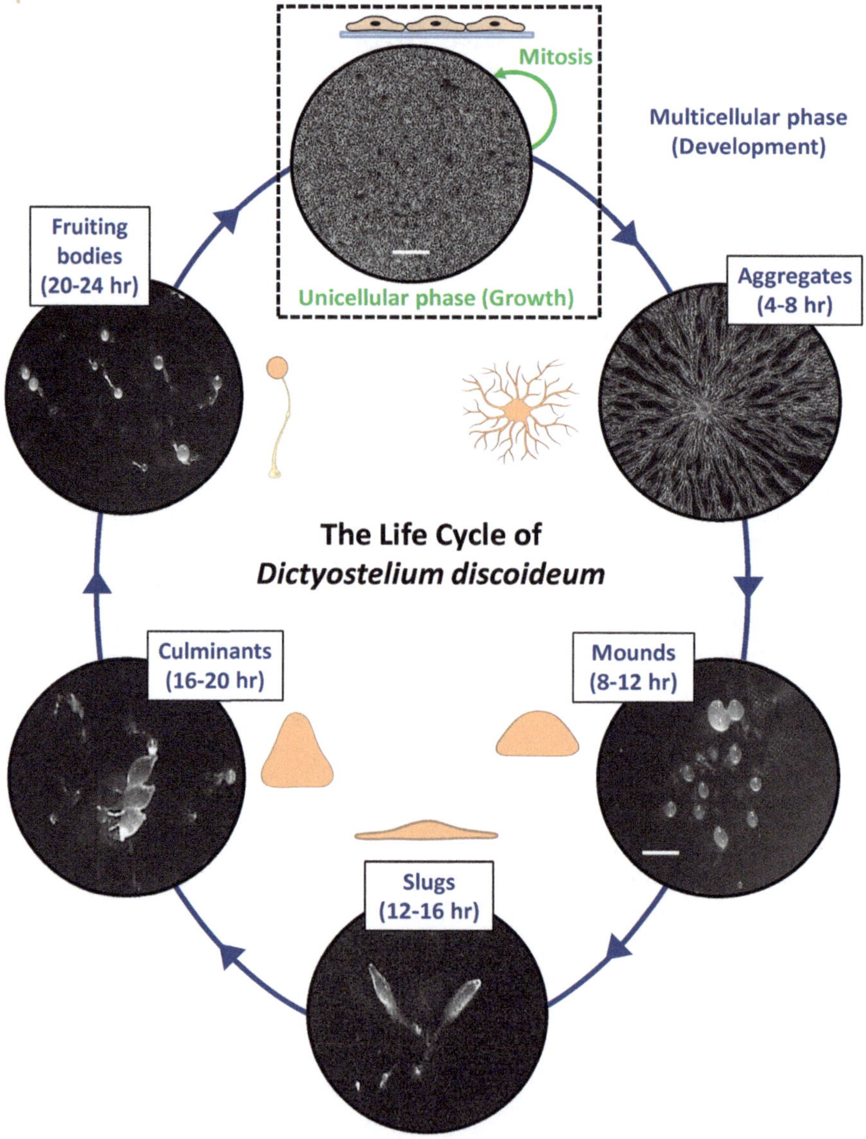

Fig. 1 The *D. discoideum* life cycle. During the unicellular growth phase, haploid amoebae feed on bacteria (e.g., *Klebsiella aerogenes*) or nutrients within axenic medium and divide mitotically. *D. discoideum* cells initiate multicellular development when nutrients are depleted. Cells aggregate into a single mound and undergo a series of developmental stages that result in the formation of a fruiting body composed of a stalk that suspends spores above the surface. Spores germinate in the presence of nutrients and restart the life cycle. Scale bar (for unicellular phase and aggregate image) = 150 μm. Scale bar for multicellular development (from mounds to fruiting bodies) = 2000 μm. (Cartoon images obtained from BioRender)

nutrient source (e.g., bacteria in nature; nutrient-rich media in the laboratory) and divide mitotically. When amoebae are starved of nutrients, the multicellular phase of the life cycle is initiated. Amoebae aggregate into a mound, which develops into a slug. Cells within the slug then terminally differentiate to form a fruiting body composed of a mass of spores that sit atop a slender stalk. When placed in a nutrient-rich environment, spores germinate and restart the life cycle.

In the laboratory, *D. discoideum* cells are inexpensive to culture and double via mitosis every 8–10 h [1]. *D. discoideum* also displays metazoan-like cellular behavior, which has facilitated its use as a model system for studying diseases caused by aberrant lysosomal function, such as neurodegeneration [2]. As a result, findings from *D. discoideum* are often translatable to more complex eukaryotic models [3–6].

1.2 Lysosomal Biology in D. discoideum

As in humans, lysosomal proteins in *D. discoideum* are synthesized and undergo post-translational modifications (e.g., N-linked glycosylation) in the endoplasmic reticulum (ER) [7]. In the ER, lysosomal proteins bind to receptors that traffic the proteins to the Golgi complex [8, 9]. Proteins are trafficked to lysosomes via mannose-6-phosphate (M6P) receptors or M6P-independent pathways, such as those involving sortilin [10, 11]. Lysosomal proteins can also be released into the extracellular space via mechanisms that are slowly being revealed [11–13]. Intriguingly, emerging evidence suggests that altered protein secretion contributes to neurodegeneration [13]. In addition, aberrant protein release during neurodegeneration can facilitate the identification of biomarkers for diagnostic or prognostic purposes, as well as providing a means to assess therapy response [14].

In *D. discoideum*, protein trafficking and secretion has been extensively studied [15–20]. During growth and aggregation, *D. discoideum* cells secrete precursor and mature forms of lysosomal enzymes, including, but not limited to, β-glucosidase, α-mannosidase, and N-acetylglucosaminidase [15, 16, 21, 22]. The release of enzymes in *D. discoideum* is an active process, and the detection of enzymes outside cells is not due to their passive release or incidental release following the egestion of intracellular material [16]. Intracellular and extracellular pools of lysosomal enzymes are sorted at the Golgi complex, and protease activity is involved in trafficking enzymes to lysosomes [21, 23]. The detection of lysosomal enzymes outside cells suggests they may function extracellularly. Indeed, previous work has detected lysosomal enzyme activity outside *D. discoideum* cells, which is thought to modulate cell adhesion and migration during the early stages of multicellular development [15, 24].

1.3 D. discoideum as a Model System for Studying Lysosomal Function

Genetic mutations in lysosomal proteins can cause neurodegenerative lysosomal storage diseases such as Pompe's disease, Fabry's disease, and the neuronal ceroid lipofuscinoses (NCLs) [25, 26]. *D. discoideum* is an excellent biomedical model for studying these and other lysosomal diseases since its genome encodes homologs for several human lysosomal proteins, including those that are absent in other early eukaryotic models (Table 1) [2, 27]. Since *D. discoideum* is haploid and its genome is sequenced and annotated, genes can be mutated, knocked down, or knocked out using several well-established approaches, including restriction enzyme-mediated integration (REMI) mutagenesis, RNA interference, homologous recombination, CRISPR interference, and CRISPR/Cas9-mediated gene editing [28–32]. Additionally, a variety of expression constructs have been generated that allow researchers to examine the localization of a protein in live or fixed cells using a fluorescent (e.g., GFP) or nonfluorescent tag (e.g., FLAG), respectively [33, 34]. Protocols for examining the autophagy pathway and autophagic flux are also well established in *D. discoideum* [35]. Moreover, the mechanisms regulating endocytosis (both fluid and solid phase) and endocytic trafficking have been well studied [36]. Finally, activity assays for conserved lysosomal enzymes are firmly established in *D. discoideum* and described in detail elsewhere [37]. In this chapter, we describe methods used to examine the intracellular localization, trafficking, and release of conserved lysosomal proteins in *D. discoideum*.

2 General Methods for Culturing *D. discoideum*

2.1 Materials

Wild-type (WT) cells, mutant cell lines, and bacteria (e.g., *Klebsiella aerogenes, Escherichia coli*) can be purchased from the Dicty Stock Center (dictybase.org) [38]. SM agar (catalog #: SMA0102) and HL5 medium containing glucose (catalog #: HLG0102) can be purchased from Formedium (Hunstanton, Norfolk, United Kingdom). Ampicillin, streptomycin sulfate, blasticidin S hydrochloride, and G418 sulfate (Geneticin) can be purchased from any chemical supplier.

2.2 Methods

Detailed methods for culturing *D. discoideum* cells have been described elsewhere [39]. Briefly, cell lines are maintained on SM agar with bacteria (e.g., *Klebsiella aerogenes* or *Escherichia coli*) at 22 °C. Cells can also be grown axenically in HL5 medium at 22 °C and 150 rpm. Cells grown in suspension are supplemented with ampicillin (100 μg/mL) and streptomycin sulfate (300 μg/mL) to prevent bacterial growth. Blasticidin S hydrochloride (10 μg/mL) is used to select and maintain gene knockout clones, and G418 sulfate (10 μg/mL) is used to maintain the expression of extrachromosomal plasmids within cells. Cells used for experiments should

Table 1
List of proteins associated with lysosomal storage diseases in humans that are conserved in *D. discoideum*

Human disease	Protein	Gene	Uniprot ID	*D. discoideum* homolog(s)	dictyBase protein ID
Aspartylglucosaminuria	N(4)-(beta-N-acetylglucosaminyl)-L-asparaginase	*AGA*	P20933	Uncharacterized protein DDB0230992	DDB0230992
Cystinosis	Cystinosin	*CTNS*	O60931	Ctns	DDB0237502
Galactosialidosis	Cathepsin A	*CTSA*	P10619	Uncharacterized protein DDB0266717 Uncharacterized protein DDB0229903 Uncharacterized protein DDB0266715	DDB0266717 DDB0229903 DDB0266715
Pycnodysostosis	Cathepsin K	*CTSK*	P43235	CprC	DDB0220784
Fucosidosis	Tissue alpha-L-fucosidase	*FUCA1*	P04066	AlfA Uncharacterized protein DDB0306858	DDB0185016 DDB0306858
Pompe disease	Lysosomal alpha-glucosidase	*GAA*	P10253	Gaa ModA	DDB0237578 DDB0191113
Gaucher disease	Lysosomal acid glucosylceramidase	*GBA1*	P04062	GluA[a] Uncharacterized protein DDB0306757[b]	DDB0215373 DDB0306757
Fabry disease	Alpha-galactosidase A	*GLA*	P06280	Uncharacterized protein DDB0233869 MelA	DDB0233869 DDB0233870
Mucopolysaccharidosis type IVB (Morquio syndrome type B)	Beta-galactosidase	*GLB1*	P16278	Glb1	DDB0266380
Mucolipidosis type II (I-cell disease)	N-acetylglucosamine-1-phosphotransferase subunits alpha/beta	*GNPTAB*	Q3T906	Gpt1	DDB0304518

(continued)

Table 1
(continued)

Human disease	Protein	Gene	Uniprot ID	*D. discoideum* homolog(s)	dictyBase protein ID
Tay–Sachs disease	Beta-hexosaminidase subunit alpha	*HEXA*	P06865	NagA NagB NagC NagD	DDB0191256 DDB0304517 DDB0304520 DDB0304516
Sandhoff disease	Beta-hexosaminidase subunit beta	*HEXB*	P07686	NagA NagB NagC NagD	DDB0191256 DDB0304517 DDB0304520 DDB0304516
Mucopolysaccharidosis type IIIC (Sanfilippo syndrome)	Heparan-alpha-glucosaminide N-acetyltransferase	*HGSNAT*	Q68CP4	Uncharacterized protein DDB0234045	DDB0234045
Wolman disease	Lysosomal acid lipase/cholesteryl ester hydrolase	*LIPA*	P38571	Lip1 Lip2 Lip3 Lip4	DDB0306185 DDB0230058 DDB0306202 DDB0230059
Alpha-mannosidosis	Lysosomal alpha-mannosidase	*MAN2B1*	O00754	ManA ManB ManC ManD ManE ManF	DDB0201569 DDB0231612 DDB0231613 DDB0231614 DDB0231615 DDB0231616
Mucolipidosis type IV	Mucolipin-1	*MCOLN1*	Q9GZU1	Mcln	DDB0233084
Schindler disease	Alpha-N-acetylgalactosaminidase	*NAGA*	P17050	Uncharacterized protein DDB0233869 MelA	DDB0233869 DDB0233870
Mucopolysaccharidosis type IIIB (Sanfilippo syndrome)	Alpha-N-acetylglucosaminidase	*NAGLU*	P54802	Naglu	DDB0238329

Niemann–Pick disease type C	NPC intracellular cholesterol transporter 1	*NPC1*	O15118	NpcA NpcB	DDB0191127 DDB0305751
	NPC intracellular cholesterol transporter 2	*NPC2*	P61916	Uncharacterized protein DDB0230194 Uncharacterized protein DDB0305562	DDB0230194 DDB0305562
Salla disease	Sialin	*SLC17A5*	Q9NRA2	Uncharacterized protein DDB0305829	DDB0305829
Niemann–Pick disease types A and B	Sphingomyelin phosphodiesterase	*SMPD1*	P17405	SgmB	DDB0232049
Proteins associated with neuronal ceroid lipofuscinosis					
CLN1 disease (NCL)	Palmitoyl protein thioesterase 1	*PPT1*	P50897	Ppt1	DDB0233890
CLN2 disease (NCL)	Tripeptidyl peptidase 1	*TPP1*	O14773	Tpp1A Tpp1B Tpp1C Tpp1D Tpp1E Tpp1F	DDB0234303 DDB0306176 DDB0185020 DDB0309140 DDB0304592 DDB0214912
CLN3 disease (NCL)	Ceroid lipofuscinosis neuronal 3	*CLN3*	Q13286	Cln3	DDB0233983
CLN4 disease (NCL)	DnaJ homolog subfamily C member 5	*DNAJC5*	Q9H3Z4	Uncharacterized protein DDB0306688 Ddj1	DDB0306688 DDB0215016
CLN5 disease (NCL)	Ceroid lipofuscinosis neuronal 5	*CLN5*	O75503	Cln5	DDB0234077
CLN7 disease (NCL)	Major facilitator superfamily domain-containing protein 8	*MFSD8*	Q8NHS3	Mfsd8	DDB0307149
CLN10 disease (NCL)	Cathepsin D	*CTSD*	P07339	CtsD	DDB0215012
CLN11 disease (NCL)	Progranulin	*GRN*	P28799	Grn	DDB0238428
CLN12 disease (NCL)	Polyamine-transporting ATPase 13A2	*ATP13A2*	Q9NQ11	Kil2	DDB0237611

(continued)

Table 1
(continued)

Human disease	Protein	Gene	Uniprot ID	*D. discoideum* homolog(s)	dictyBase protein ID
CLN13 disease (NCL)	Cathepsin F	*CTSF*	Q9UBX1	CprA	DDB0201647
				CprB	DDB0214998
				CprE	DDB0185092
				DDB0252831	DDB0252831
CLN14 disease (NCL)	BTB/POZ domain-containing protein KCTD7	*KCTD7*	Q96MP8	Uncharacterized protein DDB0347398	DDB0347398
				Uncharacterized protein DDB0238663	DDB0238663
				Kctd9	DDB0231824

[a] *gluA* encodes beta-glucosidase (821 amino acids) in *D. discoideum*

[b] *DDB_G0292446* encodes uncharacterized protein DDB0306757, which is a putative non-lysosomal glucosylceramidase (1302 amino acids) in *D. discoideum*

be in the mid-log phase of growth (1–5 × 10^6 cells/mL). Cells are starved via incubation in KK2 buffer (2.2 g/L KH_2PO_4 and 0.7 g/L K_2HPO_4, pH 6.5).

3 Examining the Intracellular Localization of Lysosomal Proteins in *D. discoideum*

3.1 Materials

Low fluorescence (LoFlo) medium (catalog #: LF1001) can be purchased from Formedium (Hunstanton, Norfolk, United Kingdom). Wheat germ agglutinin (WGA) linked to a fluorophore can be purchased from Fisher Scientific. Several different fluorophore options are available (e.g., catalog #: W7024). Mouse monoclonal anti-calnexin (catalog #: 270-390-2), mouse monoclonal anti-calreticulin (catalog #: 252-234-2), mouse monoclonal anti-V-ATPase subunit C (VatC) (catalog #: 224-256-2), mouse monoclonal anti-p80 (catalog #: H161), mouse monoclonal anti-cysteine protease G (CprG) (catalog #: H36), and mouse monoclonal anti-contact site A glycoprotein (CsaA) (catalog #: 33-294-17) can be purchased from the Developmental Studies Hybridoma Bank (University of Iowa, Iowa City, Iowa, USA). Alexa Fluor-conjugated secondary antibodies can be purchased from Fisher Scientific. Several different fluorophore options are available (e.g., catalog #: A11008). Paraformaldehyde (catalog #: 43368) can be purchased from Fisher Scientific. Prolong Gold Anti-Fade Reagent with DAPI (catalog #: P36931) is available for purchase from Fisher Scientific. Immunofluorescence (IF) blocking buffer is composed of the following reagents made in KK2 buffer: 0.1% (v/v) Triton-X100 and 0.2% (w/v) gelatin, both of which can be purchased from any chemical supplier. A tweezer and clear nail polish can be purchased from any commercial vendor.

3.2 Methods

The protocol described below can be used to assess the intracellular localization of any lysosomal protein and is adapted from methods described elsewhere (Fig. 2) [40]. In our past studies, we examined the intracellular localization of the lysosomal proteins Cln5 and cathepsin D using custom in-house antibodies and GFP-fusion constructs [11, 41–43]. However, since commercial antibodies against *D. discoideum* proteins are limited, the protein of interest can also be fused to a tag in an expression vector, which is then introduced into *D. discoideum* cells. A commercial antibody against the tag can then be used to detect the fusion protein. A variety of expression vectors incorporating different tags (e.g., GFP, FLAG, Myc, etc.) are available from the Dicty Stock Center [33, 34, 38]. Protocols for generating expression constructs and transforming them into *D. discoideum* cells are well established and described in detail elsewhere [44].

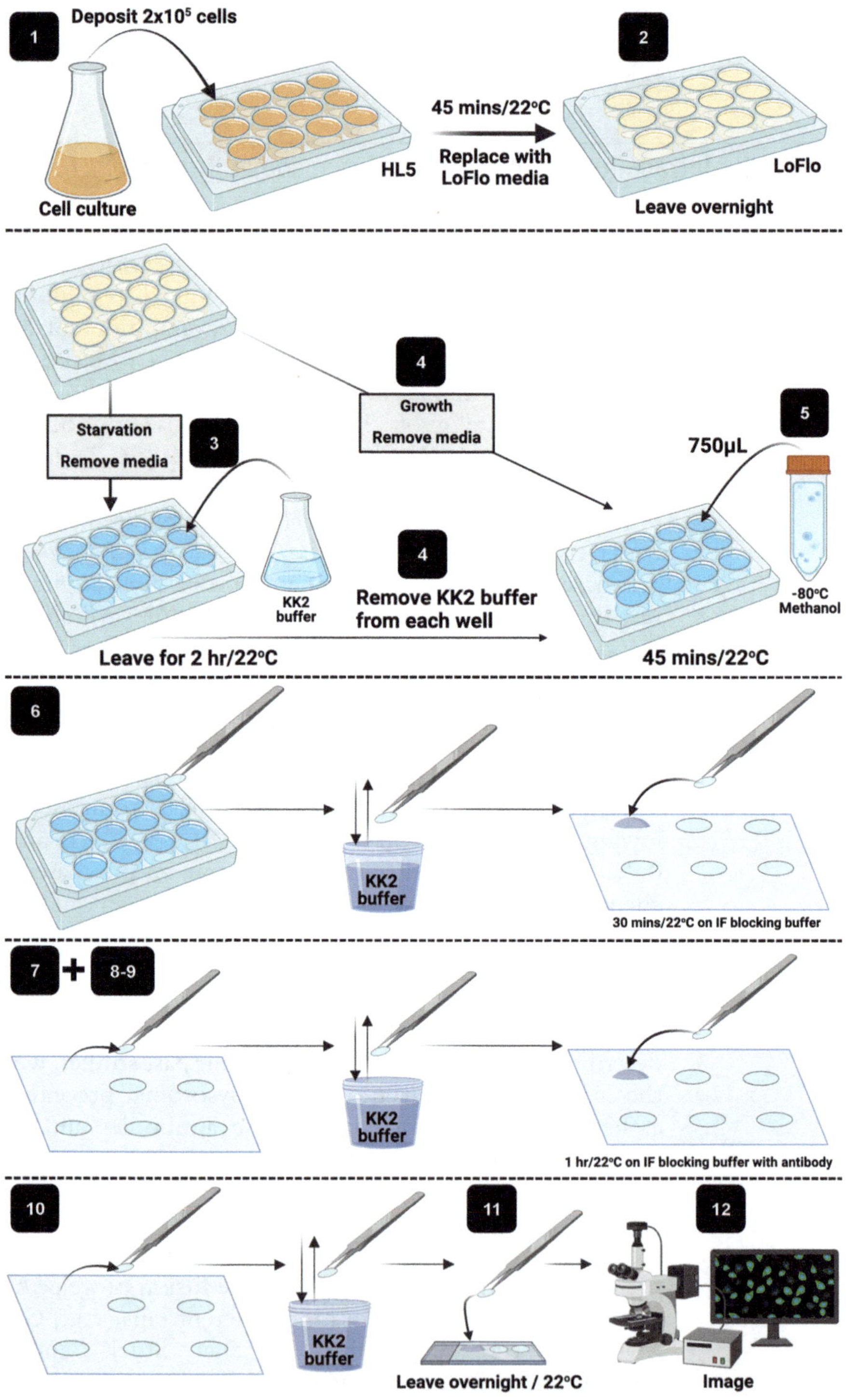

Fig. 2 Immunofluorescence assay (1). 2×10^5 cells in the mid-log phase of growth are deposited onto coverslips placed inside separate wells of a 12-well dish containing HL5 medium. Once cells are adhered to the coverslip (45 min), (2) the HL5 medium is replaced with low fluorescence (LoFlo) medium, and the dish is left overnight at 22 °C (3). To examine localization during starvation, coverslips are washed twice with 750 μL of fresh KK2 buffer and then submerged in 750 μL of KK2 buffer for 2 h (4). For both growth and starvation, the LoFlo medium/KK2 buffer is removed, and (5) 750 μL of −80 °C methanol is added to each well and left for

1. 2×10^5 cells in the mid-log phase of growth are deposited into a 12-well dish containing HL5 medium, the appropriate antibiotics, and a sterile coverslip.
2. Cells are left to adhere on the coverslip for at least 1 h at 22 °C, after which time the medium is replaced with LoFlo medium containing the appropriate antibiotics. Cells are left overnight at 22 °C to double twice to a confluency of no more than 80% (*see* **Note 1**).
3. For fixing and imaging growth-phase cells, proceed to **step 4** (*see* **Note 2**). To assess protein localization during starvation, the LoFlo medium in the well is removed, cells on the coverslip are washed twice with 750 μL KK2 buffer, and the coverslip is incubated in 750 μL of KK2 buffer for 2 h (*see* **Notes 2** and **3**).
4. Following incubation, LoFlo medium (growth) or KK2 buffer (starvation) is removed from the well, and 750 μL of −80 °C methanol is gently dispensed against the side of the well (*see* **Note 4**).
5. Methanol is left in the well for 45 min at 22 °C, after which time the coverslip is removed from the well with a tweezer, rinsed thoroughly by dunking 20 times in a reservoir filled with KK2 buffer, and placed cell-side down onto a 40-μL droplet of IF blocking buffer on parafilm for 30 min at 22 °C (*see* **Note 5**).
6. Following incubation in IF blocking buffer, the edge of the coverslip is dabbed on a paper towel and then dunked 20 times in a fresh reservoir containing KK2 buffer. The coverslip is then placed onto a 40-μL droplet of IF blocking buffer containing primary antibody (see below) that targets an organelle in the lysosomal protein trafficking pathway (*see* **Note 6**) and left to incubate for 1 h at 22 °C.
   - Anti-calnexin and anti-calreticulin target the ER.
   - WGA targets the Golgi complex (*see* **Note 2**).
   - Anti-VatC targets acidic vesicles such as lysosomes.
   - Anti-p80 targets secretory lysosomes.
   - Anti-CprG targets multivesicular bodies.
   - Anti-CsaA targets the cell periphery.

Fig. 2 (continued) 45 min at 22 °C (6). Coverslips are rinsed in KK2 buffer and then placed cell-side down onto 40 μL of immunofluorescence (IF) blocking buffer for 30 min at 22 °C (7). Coverslips are rinsed in KK2 buffer and placed cell-side down onto 40 μL of IF blocking buffer containing a primary antibody targeting the protein of interest for 1 h at 22 °C (8 and 9). The previous step is repeated for the primary antibody targeting the marker, and then again for the secondary antibodies targeting the two primary antibodies (10). Once the coverslips are rinsed in KK2 buffer and (11) placed onto a microscope slide with mounting solution containing DAPI, (12) the cells on the coverslips are imaged using a fluorescence microscope. (Figure generated using BioRender)

7. Following incubation with the first primary antibody, the coverslip is dunked 20 times in a fresh reservoir of KK2 buffer and then placed cell-side down onto a 40 μL droplet of IF blocking buffer containing the second primary antibody, which targets the protein of interest or the tag (if a fusion protein is being studied). The coverslip is incubated in the primary antibody for 1 h at 22 °C.
8. Following incubation with the second primary antibody, the coverslip is dunked 20 times in a fresh reservoir of KK2 buffer. The coverslip is then placed cell-side down onto a 40-μL droplet of IF blocking buffer containing two secondary antibodies that each target one of the primary antibodies (*see* **Note 7**). The coverslip is incubated with the secondary antibodies for 1 h at 22 °C.
9. Following incubation with the secondary antibodies, the coverslip is dunked 20 times in a fresh reservoir of KK2 buffer and then placed cell-side down onto a drop of Prolong Gold Anti-Fade Reagent with DAPI on a microscope slide.
10. Slides are incubated overnight at 22 °C and protected from light. The following day, clear nail polish can be used to seal the edge of the coverslip to the slide.
11. Images of cells are taken with a standard fluorescence microscope equipped with the appropriate filters to capture the wavelength of light emitted by the fluorophore-linked secondary antibodies.

3.3 Data Analysis

Multiple coverslips should be imaged to determine the localization of the protein of interest, the effect of gene mutation, or the impact of a chemical treatment on protein localization. If quantitative analyses are desired, it is important to standardize the exposure times across all images. The arbitrary fluorescent unit can be measured using Fiji/ImageJ [45], and the arbitrary fluorescent unit values can be standardized to the exposure time. The arbitrary fluorescent unit/exposure time values from the knockout or treatment group are then normalized to the WT or control group. Finally, a one-sample t-test with a p-value of <0.05 can be performed to determine statistical significance.

3.4 Notes

1. A chemical such as chloroquine (autophagy inhibitor) or latrunculin A (actin polymerization inhibitor) can be added at this step to study its effect on the localization of the protein of interest during growth.
2. To examine co-localization to the Golgi complex, a final concentration of 1 μg/mL WGA conjugated to a fluorophore is added into either the LoFlo medium (growth) or KK2 buffer (starvation) for 30 min at 22 °C, along with two washing steps with either LoFlo medium (growth) or KK2 buffer (starvation) [19].

3. A chemical such as chloroquine (autophagy inhibitor) or latrunculin A (actin polymerization inhibitor) can be added at this step to study its effect on the localization of the protein of interest during starvation. A 2 h starvation period is recommended since many cells de-adhere from coverslips after longer starvation times. Finally, it is recommended that the KK2 buffer be gently dispensed against the interior wall of the well to prevent cells from de-adhering from the coverslip.
4. In this step, paraformaldehyde fixation can be used as an alternative method if issues are encountered with methanol fixation. Paraformaldehyde fixation involves incubating cells attached to coverslips in 750 μL of 4% paraformaldehyde (v/v) in growth medium or KK2 buffer for 20 min, washing the coverslip with KK2 buffer, submerging the coverslips in −20 °C methanol for 5 min, washing the coverslip with KK2 buffer, and then finally placing the coverslip cell-side down onto IF blocking buffer.
5. Ensure that the parafilm is large enough so that the 12-well dish lid creates a seal to hold the moisture in. This will prevent the droplets from evaporating.
6. Typically, a primary antibody dilution of 1:50 or 1:100 works best. However, researchers should test out a series of dilutions to determine the optimal one to use.
7. Typically, a secondary antibody dilution of either 1:100 or 1: 200 works best. Ensure that the two primary antibodies used in the experiment are developed in different animal species. This will allow the selection of two secondary antibodies with different fluorophores that each target the appropriate antibody species. For example, anti-VatC (noted above) is made in mice, whereas the antibody you use to target your protein of interest may be made in rabbits. In this situation, you would use donkey anti-mouse Alexa 555 (red color) to reveal VatC intracellular localization and goat anti-rabbit Alexa 488 (green color) to reveal your protein of interest. If required, other secondary antibody options are available commercially (e.g., different host species and different fluorophores).

4 Examining the Release of Lysosomal Proteins from *D. discoideum* Cells

4.1 Materials

Protease cocktail inhibitor tablets (catalog #: PIA32955) can be purchased from Fisher Scientific. Qubit 2.0 fluorometer can be purchased from Fisher Scientific. Mouse monoclonal anti–αactinin (catalog #: 47-18-9), mouse monoclonal anti-β-actin (catalog #: 224-236-1), and mouse monoclonal anti-α-tubulin (catalog #: 12G10) can be purchased from the Developmental Studies

Hybridoma Bank (University of Iowa, Iowa City, Iowa, USA). Goat anti-rabbit IgG linked to HRP (catalog #: 7074) and horse anti-mouse IgG linked to HRP (catalog #: 7076) can be purchased from Cell Signaling Technology. Chemicals that affect lysosomal function can be purchased from commercial vendors: ammonium chloride (Fisher Scientific, catalog #: A661-500), bafilomycin A1 (Fisher Scientific, catalog #: AAJ61835MCR), chloroquine diphosphate salt (Sigma Aldrich, catalog #: C6628), and vacuolin-1 (Sigma Aldrich, catalog #: 673000). Amicon Ultra-4 Centrifugal Filter units can be purchased from Sigma Aldrich (catalog #: UFC801024). Clarity Max™ Western Enhanced Chemiluminescence (ECL) Substrate solution (catalog #: 1705062) and a ChemiDoc Imaging System device can be purchased from Bio-Rad Laboratories Limited. 2× SDS sample buffer is composed of the following reagents, which can be purchased from any chemical supplier: 4% (w/v) sodium dodecyl sulfate, 10% (v/v) beta-mercaptoethanol, 0.125 M Tris, 20% (v/v) glycerol, bromophenol blue, and pH 6.8. Lysis buffer is composed of the following reagents, which can also be purchased from any chemical supplier: 150 mM NaCl, 50 mM Tris, 0.5% (v/v) NP-40, and pH 8.3.

4.2 Methods

The protocol described below can be used to assess the intracellular and extracellular amounts of any lysosomal protein, provided that an antibody against the protein is available (Fig. 3). In our past studies, we examined the intracellular and extracellular amounts of the lysosomal proteins Cln5 and cathepsin D using custom in-house antibodies [42, 43]. As noted in Subheading 3.2, cell lines expressing a fusion protein (e.g., GFP, FLAG, Myc, etc.) and a primary antibody directed against the tag can be used to circumvent the limited availability of commercial antibodies against *D. discoideum* proteins [33, 34]. Furthermore, mutants can be purchased from the Dicty Stock Center to determine pathways that regulate the trafficking of lysosomal proteins and their release from cells [11, 38]. Finally, a 4 h starvation period is recommended since this is the point during the life cycle when enzymes are maximally secreted [15, 16].

1. 5 × 10^6 cells in the mid-log phase of growth are deposited into 100 × 15 mm Petri dishes and left to adhere to the dish for at least 45 min at 22 °C.
2. Once the cells are adhered to the dish, the medium is replaced with 10 mL of fresh HL5 medium containing the appropriate antibiotics and incubated overnight at 22 °C (*see* **Note 1**).
3. The following day, confluent cells are washed twice with 10 mL of KK2 buffer and then submerged in 10 mL of fresh KK2 buffer for 4 h (*see* **Note 2**).

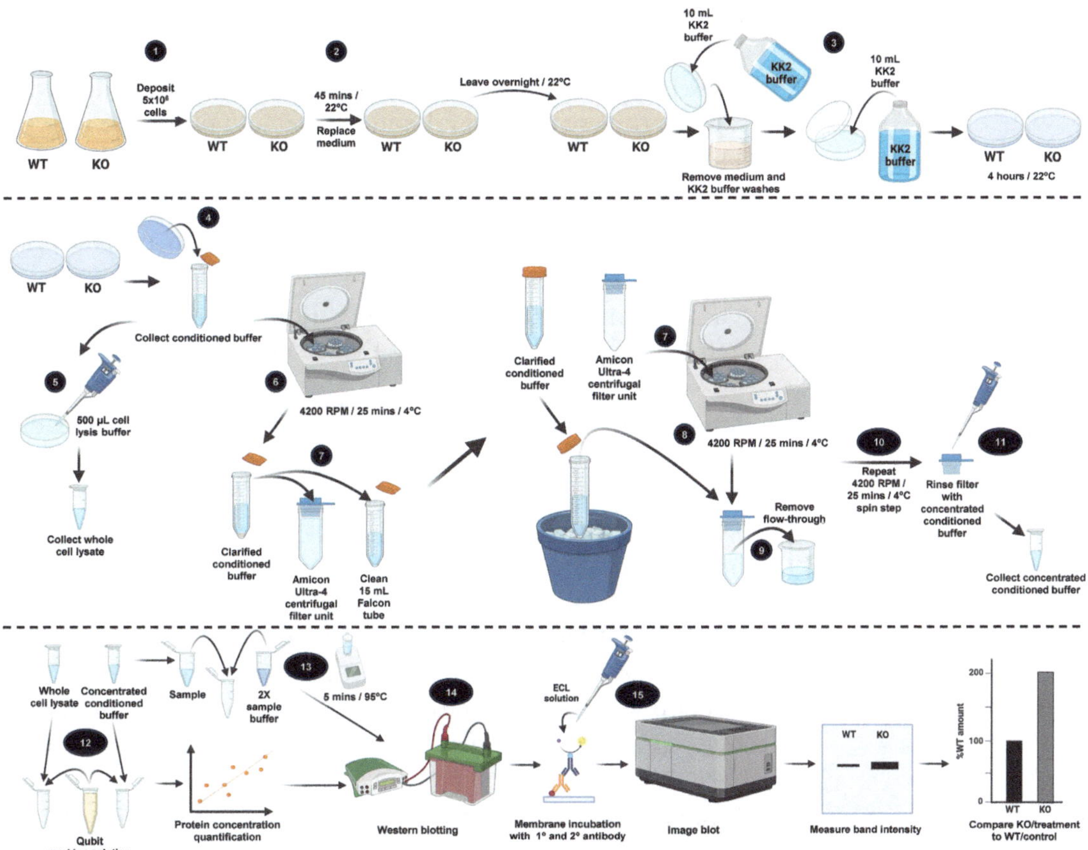

Fig. 3 Secretion assay (1). 5×10^6 cells in the mid-log phase of growth are deposited onto 100×15 mm Petri dishes with HL5 medium and left for 45 min at 22 °C. Once the cells are adhered to the Petri dish, (2) the HL5 medium is replaced with fresh HL5 medium containing the appropriate antibiotics and left overnight at 22 °C (3). The following day, the cells on the Petri dish are washed twice with 10 mL of fresh KK2 buffer and submerged in 10 mL of KK2 buffer for 4 h (4). The conditioned buffer is collected into a 15 mL Falcon tube, and (5) the starved cells are lysed with 500 μL of lysis buffer containing a dissolved protease inhibitor tablet and then collected into a 1.5 mL tube (6). The conditioned buffer is centrifuged at 2500 RPM for 5 min at 4 °C to pellet any free-floating cells (7). The clarified conditioned buffer is collected into a sterile 15 mL Falcon tube, and 4 mL of clarified conditioned buffer is added into an Amicon Ultra-4 Centrifugal Filter unit (8). The Amicon Ultra-4 Centrifugal Filter unit is centrifuged at 4200 RPM for 25 min at 4 °C, (9) the flow-through is discarded, and the clarified conditioned buffer is reloaded into the Amicon Ultra-4 Centrifugal Filter unit (10). The previous step is repeated until all the clarified conditioned buffer is concentrated in the Amicon Ultra-4 Centrifugal Filter unit (11). The concentrated conditioned buffer is used to rinse the filters in the Amicon Ultra-4 Centrifugal Filter unit and then collected into a 1.5 mL tube (12). The protein concentration in whole cell lysates and concentrated conditioned buffer is quantified (13). Equal volumes of whole cell lysate or concentrated conditioned buffer are mixed with 2× sample buffer. Solutions are then heated for 5 min at 95 °C (14). Proteins in the mixture are separated by SDS-PAGE and analyzed by western blotting using standard methods (15). Enhanced chemiluminescence (ECL) substrate solution is pipetted onto membranes containing primary and secondary antibodies and imaged using a device such as the Bio-Rad ChemiDoc. Finally, the band intensity for each protein is measured with Fiji/ImageJ, standardized against a loading control (for intracellular samples only: α-actinin, β-actin, or α-tubulin), and knockout or treatment values are further standardized with respect to the WT or control group, respectively. (Figure generated using BioRender)

4. After 4 h of starvation, the conditioned buffer is collected into a 15 mL centrifuge tube placed on ice.
5. Cells in the dish are lysed with lysis buffer containing a dissolved protease cocktail inhibitor tablet and placed on ice.
6. The conditioned buffer is spun in a centrifuge at 2500 RPM at 4 °C for 5 min to pellet free-floating cells.
7. The clarified conditioned buffer is collected into a new 15 mL centrifuge tube and placed on ice.
8. The proteins in the clarified conditioned buffer are concentrated using an Amicon Ultra-4 Centrifugal Filter unit at 4200 RPM and 4 °C for 25 min.
9. After 25 min, the flow-through in the collection tube is discarded, and the Amicon Ultra-4 Centrifugal Filter unit is placed back into the collection tube.
10. **Steps 8** and **9** are repeated until all the clarified conditioned buffer is concentrated in the Amicon Ultra-4 Centrifugal Filter unit. A total of 2–3 spins may be required, depending on how much buffer needs to be concentrated (*see* **Note 3**).
11. Following centrifugation, both sides of the interior filter surface are washed five times with the concentrated conditioned buffer solution, which is then collected into a 1.5 mL tube placed on ice.
12. Protein amounts within whole cell lysates and samples of concentrated conditioned buffer are quantified using a Qubit 2.0 Fluorometer according to the manufacturer's instructions (*see* **Note 4**).
13. Proteins in whole cell lysates and concentrated conditioned buffer are denatured in a 1:1 v/v mixture of 2× sample buffer and sample, and then heated at 95 °C for 5 min.
14. Proteins within prepared samples are separated by SDS-PAGE and analyzed by western blotting using standard methods (2-h incubation at 22 °C for primary and secondary antibodies in 5% milk/TBST) [46]. In our experience, wet transfer works best.
15. Protein bands are imaged using the Clarity Max ECL substrate solution and the ChemiDoc Imaging System (*see* **Note 5**). Mouse monoclonal anti-α-actinin, mouse monoclonal anti-β-actin, and mouse monoclonal anti-α-tubulin are used as loading controls for intracellular protein quantification.

4.3 Data Analysis

Protein bands are quantified using Fiji/ImageJ [45]. For intracellular measures, the quantified protein bands are standardized against either α-actinin, β-actin, or α-tubulin. Equal protein loading is the only method for controlling for extracellular protein amounts since there is no established extracellular control for *D. discoideum* (*see* **Note 6**). Values from the mutant line or a treatment group are expressed as a percentage of the value in the WT line or control group, respectively. Finally, a one-sample t-test with a p-value of <0.05 is performed to determine statistical significance.

4.4 Notes

1. A master mix containing the medium and a chemical, such as chloroquine (autophagy inhibitor) or latrunculin A (actin polymerization inhibitor), can be prepared to study the effect of the chemical on protein secretion during growth.
2. A master mix containing the medium and a chemical, such as chloroquine (autophagy inhibitor) or latrunculin A (actin polymerization inhibitor), can be prepared to study the effect of the chemical on protein secretion during starvation.
3. Larger Amicon Ultra-4 Centrifugal Filter units (Sigma Aldrich, catalog #: UFC9010) can be used for this step if only one spin is desired or larger volumes of conditioned buffer (e.g., >10 mL) need to be concentrated.
4. Any protein quantification method can be used as an alternative (e.g., BCA assay, Bradford assay).
5. Any ECL substrate solution, as well as any ECL gel doc imaging equipment, can be used.
6. Silver-staining the gel or Ponceau staining the membrane prior to the blocking step can be done to ensure equal extracellular protein amounts.

References

1. Mathavarajah S, Flores A, Huber RJ (2017) *Dictyostelium discoideum*: a model system for cell and developmental biology. Curr Protoc Essent Lab Tech 15:14.1.1–14.1.19
2. Haver HN, Scaglione KM (2021) *Dictyostelium discoideum* as a model for investigating neurodegenerative diseases. Front Cell Neurosci 15:1–18
3. Williams RSB, Boeckeler K, Gräf R, Müller-Taubenberger A, Li Z, Isberg RR, Wessels D, Soll DR, Alexander H, Alexander S (2006) Towards a molecular understanding of human diseases using *Dictyostelium discoideum*. Trends Mol Med 12:415–424
4. Alexander S, Alexander H (2011) Lead genetic studies in *Dictyostelium discoideum* and translational studies in human cells demonstrate that sphingolipids are key regulators of sensitivity to cisplatin and other anticancer drugs. Semin Cell Dev Biol 22:97–104
5. Huber RJ (2020) Molecular networking in the neuronal ceroid lipofuscinoses: insights from mammalian models and the social amoeba *Dictyostelium discoideum*. J Biomed Sci 27:64
6. Kroll J, Hauschild R, Kuznetcov A, Stefanowski K, Hermann MD, Merrin J, Shafeek L, Müller-Taubenberger A, Renkawitz J (2023) Adaptive pathfinding by nucleokinesis during amoeboid migration. EMBO J 42:1–18
7. Cardelli JA, Golumbeski GS, Dimond RL (1986) Lysosomal enzymes in *Dictyostelium discoideum* are transported to lysosomes at distinctly different rates. J Cell Biol 102:1264–1270
8. Knecht DA, Dimond RL (1981) Lysosomal enzymes possess a common antigenic determinant in the cellular slime mold, *Dictyostelium discoideum*. J Biol Chem 256:3564–3575
9. Cardelli JA, Mierendorf RC Jr, Dimond RL (1986) Initial events involved in the synthesis of the lysosomal enzyme alpha-mannosidase in *Dictyostelium discoideum*. Arch Biochem Biophys 244:338–345

10. Ruscetti T, Cardelli JA, Niswonger ML, O'Halloran TJ (1994) Clathrin heavy chain functions in sorting and secretion of lysosomal enzymes in *Dictyostelium discoideum*. Nature 258:273–273
11. Huber RJ, Kim WD, Wilson-Smillie MLDM (2024) Mechanisms regulating the intracellular trafficking and release of CLN5 and CTSD. Traffic 25:e12925
12. Tancini B, Buratta S, Delo F, Sagini K, Chiaradia E, Pellegrino RM, Emiliani C, Urbanelli L (2020) Lysosomal exocytosis: the extracellular role of an intracellular organelle. Membranes 10:1–21
13. Huber RJ (2021) Altered protein secretion in batten disease. Dis Model Mech 14: dmm049152
14. Herman M, Randall GW, Spiegel JL, Maldonado DJ, Simoes S (2024) Endo-lysosomal dysfunction in neurodegenerative diseases: opinion on current progress and future direction in the use of exosomes as biomarkers. Philos Trans R Soc Lond Ser B Biol Sci 379: 20220387
15. Rossomando EF, Maldonado B, Crean EV, Kollar EJ (1978) Protease secretion during onset of development in *Dictyostelium discoideum*. J Cell Sci 30:305–318
16. Dimond RL, Burns RA, Jordan KB (1981) Secretion of lysosomal enzymes in the cellular slime mold, *Dictyostelium discoideum*. J Biol Chem 256:6565–6572
17. Bakthavatsalam D, Gomer RH (2010) The secreted proteome profile of developing *Dictyostelium discoideum* cells. Proteomics 10: 2556–2559
18. Huber RJ, O'Day DH (2015) Proteomic profiling of the extracellular matrix (slime sheath) of *Dictyostelium discoideum*. Proteomics 15:3315–3319
19. Huber RJ (2017) Loss of Cln3 impacts protein secretion in the social amoeba *Dictyostelium*. Cell Signal 35:61–72
20. Huber RJ, Gray J, Kim WD (2023) Loss of *mfsd8* alters the secretome during *Dictyostelium* aggregation. Eur J Cell Biol 102:151361
21. Mierendorf RC, Cardelli JA, Livi GP, Dimond RL (1983) Synthesis of related forms of the lysosomal enzyme alpha-mannosidase in *Dictyostelium discoideum*. J Biol Chem 258: 5878–5884
22. Cardelli JA, Schatzle J, Bush JM, Richardson J, Ebert D, Freeze H (1990) Biochemical and genetic analysis of the biosynthesis, sorting, and secretion of *Dictyostelium* lysosomal enzymes. Dev Genet 11:454–462
23. Richardson JM, Woychik NA, Ebert DL, Dimond RL, Cardelli JA (1988) Inhibition of early but not late proteolytic processing events leads to the missorting and oversecretion of precursor forms of lysosomal enzymes in *Dictyostelium discoideum*. J Cell Biol 107:2097–2107
24. Huber RJ, Mathavarajah S (2019) Comparative transcriptomics reveals mechanisms underlying *cln3*-deficiency phenotypes in *Dictyostelium*. Cell Signal 58:79–90
25. Pastores GM, Maegawa GHB (2013) Clinical neurogenetics: neuropathic lysosomal storage disorders. Neurol Clin 31:1051–1071
26. Mole SE, Cotman SL (2015) Genetics of the neuronal ceroid lipofuscinoses (Batten disease). Biochim Biophys Acta Mol basis Dis 1852:2237–2241
27. Huber RJ, Hughes SM, Liu W, Morgan A, Tuxworth RI, Russell C (2020) The contribution of multicellular model organisms to neuronal ceroid lipofuscinosis research. Biochim Biophys Acta Mol basis Dis 1866:165614
28. Eichinger L, Pachebat J, Glöckner G et al (2005) The genome of the social amoeba *Dictyostelium discoideum*. Nature 435:43–57
29. Kuspa A (2006) Restriction enzyme-mediated integration (REMI) mutagenesis. Methods Mol Biol 346:201–209
30. Faix J, Linkner J, Nordholz B, Platt JL, Liao XH, Kimmel AR (2013) The application of the Cre-loxP system for generating multiple knock-out and knock-in targeted loci. Methods Mol Biol 983:249–267
31. Friedrich M, Meier D, Schuster I, Nellen W (2015) A simple retroelement based Knockdown system in *Dictyostelium*: further insights into RNA interference mechanisms. PLoS One 10:e0131271
32. Yamashita K, Iriki H, Kamimura Y, Muramoto T (2021) CRISPR toolbox for genome editing in *Dictyostelium*. Front Cell Dev Biol 9:721630
33. Levi S, Polyakov M, Egelhoff TT (2000) Green fluorescent protein and epitope tag fusion vectors for *Dictyostelium discoideum*. Plasmid 44: 231–238
34. Veltman DM, Akar G, Bosgraaf L, Van Haastert PJ (2009) A new set of small, extrachromosomal expression vectors for *Dictyostelium discoideum*. Plasmid 61:110–118
35. Domínguez-Martín E, Cardenal-Muñoz E, King JS, Soldati T, Coria R, Escalante R (2017) Methods to monitor and quantify autophagy in the social amoeba *Dictyostelium discoideum*. Cells 6:18

36. Vines JH, King JS (2019) The endocytic pathways of *Dictyostelium discoideum*. Int J Dev Biol 63:461–471
37. Kim WD, DiGiacinto AF, Huber RJ (2024) Assaying lysosomal enzyme activity in *Dictyostelium discoideum*. Methods Mol Biol 2814: 55–79
38. Fey P, Dodson RJ, Basu S, Hartline EC, Chisholm RL (2019) dictyBase and the Dicty Stock Center (version 2.0) – a progress report. Int J Dev Biol 63:563–572
39. Fey P, Kowal AS, Gaudet P, Pilcher KE, Chisholm RL (2007) Protocols for growth and development of *Dictyostelium discoideum*. Nat Protoc 2:1307–1316
40. Hagedorn M, Neuhaus EM, Soldati T (2006) Optimized fixation and immunofluorescence staining methods for *Dictyostelium* cells. Methods Mol Biol 346:327–338
41. Huber RJ, Mathavarajah S (2018) Cln5 is secreted and functions as a glycoside hydrolase in *Dictyostelium*. Cell Signal 42:236–248
42. Huber RJ, Mathavarajah S (2018) Secretion and function of Cln5 during the early stages of *Dictyostelium* development. Biochim Biophys Acta, Mol Cell Res 1865:1437–1450
43. Huber RJ, Mathavarajah S, Yap SQ (2020) Mfsd8 localizes to endocytic compartments and influences the secretion of Cln5 and cathepsin D in *Dictyostelium*. Cell Signal 70: 109572
44. Gaudet P, Pilcher KE, Fey P, Chisholm RL (2007) Transformation of *Dictyostelium discoideum* with plasmid DNA. Nat Protoc 2:1317–1324
45. Schindelin J, Arganda-Carreras I, Frise E, Kaynig V, Longair M, Pietzsch T, Preibisch S, Rueden C, Saalfeld S, Schmid B, Tinevez JY, White DJ, Hartenstein V, Eliceiri K, Tomancak P, Cardona A (2012) Fiji: an open-source platform for biological-image analysis. Nat Methods 9:676–682
46. Mahmood T, Yang PC (2012) Western blot: technique, theory, and trouble shooting. N Am J Med Sci 4:429–434

Chapter 15

Bioenergetic Profiling Applied to a Zebrafish Model of Neuronal Ceroid Lipofuscinosis, a Lysosomal Storage Disorder

Valentina Marchica, Emily Malini Young, Nastassja Ramdeen, Biggy Simbi, and Claire Russell

Abstract

Neuronal Ceroid Lipofuscinoses (NCLs) are a group of inherited neurodegenerative disorders characterized by autofluorescent lipopigment accumulation. NCLs consist of various types, spanning from CLN1 to CLN14 (except for "CLN9"), each linked to a particular mutated gene. Although the symptoms exhibit resemblances, the age at which each symptom manifests differs across the distinct types. Mitochondrial impairment has been implicated in the pathogenesis of NCLs, with studies suggesting contributions from dysfunctional oxidative phosphorylation and altered mitochondrial membrane potential to the progressive neurodegeneration observed in affected individuals. Utilizing Seahorse XF Analyzer, this study presents a method for the nuanced exploration of mitochondrial function and cellular metabolism, offering insights into potential therapeutic avenues. Zebrafish models reveal phenotypic similarities to human patients, making them promising for disease studies. Our research optimizes the Seahorse Mito Stress assay at 2 days post-fertilization (dpf) in zebrafish. This emerging approach holds great promise for identifying novel therapies.

Key words Neuronal Ceroid Lipofuscinosis, Batten Disease, Lysosomal Storage Disorder, Bioenergetics, Mitochondria, Seahorse Analyzer, Oxygen consumption rate

1 Introduction

Batten disease, also known as Neuronal Ceroid Lipofuscinoses (NCLs), comprises a group of inherited, fatal, neurodegenerative disorders characterized by the accumulation of autofluorescent lipopigments within cells [1]. Mutations in specific genes give each form of NCL its name (CLN1 to CLN14, except for "CLN9"). Within each distinct form, the majority of patients have a specific age of onset, sequence of symptoms, and rate of

Valentina Marchica, Emily Malini Young and Nastassja Ramdeen contributed equally.

Stephanie M. Hughes and Indranil Basak (eds.), *Lysosomes: Methods and Protocols*, Methods in Molecular Biology, vol. 2976,
https://doi.org/10.1007/978-1-0716-4844-5_15,

progression [2]. One of the most prevalent juvenile forms is CLN3 disease, which is caused by mutations in the *CLN3* gene [3]. CLN2 disease is the most common late infantile form [4]. Understanding the metabolic disturbances and bioenergetic alterations in the NCLs [5, 6] is crucial for advancing our knowledge of this devastating condition and developing potential therapies.

The study of bioenergetics, which involves assessing cellular energy production and consumption, has emerged as a valuable approach for uncovering the intricacies of neurodegenerative diseases like the NCLs. In this context, the Seahorse Agilent system has significantly enhanced our ability to probe the bioenergetic profiles of cells and model organisms. Cellular metabolism can be assessed through various parameters, including basal respiration, maximal respiration, proton leak, ATP production, and non-mitochondrial respiration. Basal respiration reflects the oxygen consumption rate (OCR) of cells under resting conditions, while maximal respiration represents their peak metabolic capacity. Proton leak involves the leakage of protons across the mitochondrial membrane, impacting ATP synthesis efficiency. ATP production measures the rate of ATP generation through oxidative phosphorylation. Non-mitochondrial respiration accounts for oxygen consumption not associated with ATP synthesis. Together, these measurements offer insights into cellular energetics, mitochondrial function, and metabolic regulation [7]. Mitochondrial dysfunction is evident in NCL [8], including mitochondrial ATP synthase subunit c accumulation, mitochondrial structural abnormalities, and selective neuronal loss in metabolically active brain regions in mice modeling CLN1 disease [9, 10]. In CLN1, CLN3, and CLN6 disease patients, in vitro studies confirm altered mitochondrial networks, low protein expression, and involvement of apoptosis pathways [11–14]. Impairment of mitochondrial functions is directly evident in CLN5 disease cell models and mouse cerebral cortex models [11], in astrocytes from mouse models of CLN3 disease [15], and in neurons from mouse models of CLN7 disease [12].

Unlike previous methods that may be less precise and time-consuming, the Agilent Seahorse system provides several advantages for examining the bioenergetics of the NCLs. It enables real-time monitoring of oxygen consumption and extracellular acidification in a higher-throughput manner using 96-well plates, facilitating the assessment of mitochondrial function and glycolytic activity [16]. This dynamic profiling offers a comprehensive view of cellular metabolism and its response to therapeutic interventions, making it an invaluable tool for studying the NCLs.

Zebrafish share a high degree of genetic homology with humans, making them an excellent model for exploring disease mechanisms and drug discovery. Our research employs zebrafish models involving the NCL genes, specifically modeling CLN2 and CLN3 disease [17, 18]. Remarkably, these models exhibit a range of phenotypic similarities to human patients, including impaired motor function, vision loss, and neuroinflammation. These characteristics make zebrafish a promising platform for studying NCL diseases, surpassing certain limitations of rodent models, such as their size and cost. Therefore, utilizing the zebrafish model for investigating NCLs brings distinct advantages. For example, they are the only vertebrates that can be used for in vivo bioenergetic studies using multiwell plates in the Seahorse system, enabling the testing of several conditions on one plate [19].

While the Seahorse system has revolutionized metabolic studies, its application to zebrafish is still relatively nascent. Current research has provided valuable insights into zebrafish bioenergetics. However, some limitations persist, including challenges associated with adapting the Seahorse technology to zebrafish larvae and developing well-established protocols tailored to this model organism. The existing protocols, initially designed for cells and adapted to zebrafish models, do not produce all the effects expected when mitochondrial inhibitors are added [20, 21]. This may be partly due to the difficulty in placing zebrafish in the cell culture plates. Our modified protocol employs spheroid plates, enhancing the success of zebrafish placement compared to using a cell culture plate. The assay cycles are adjusted due to the fact that zebrafish and cells require different cycling parameters. For example, we have increased the cycles to accurately observe the effect of FCCP on basal respiration curves. This protocol adaptation is discussed in the context of Seahorse instrumentation.

In our quest to unravel the metabolic intricacies of the NCLs using the zebrafish, we have harnessed the Seahorse Agilent system. Through rigorous experimentation, we have devised a highly optimized protocol for conducting the Seahorse Mito Stress assay on individual zebrafish at a critical developmental stage, 2 days post-fertilization (dpf). This refined methodology has allowed us to delve into the bioenergetics of zebrafish.

In summary, the NCLs pose a formidable challenge in the realm of neurodegenerative disorders. The Seahorse Agilent system, coupled with zebrafish models, offers a promising avenue for investigating the distinctive bioenergetic alterations in this disease and facilitates the exploration of novel therapeutic strategies. This emerging approach holds great potential for advancing our understanding of NCL and accelerating the development of effective treatments.

2 Materials

Agilent Seahorse has introduced a cutting-edge 96-well microplate specially crafted for assessing organoids. We have optimized them to assess individual 2 dpf zebrafish with the Agilent Seahorse XFe96

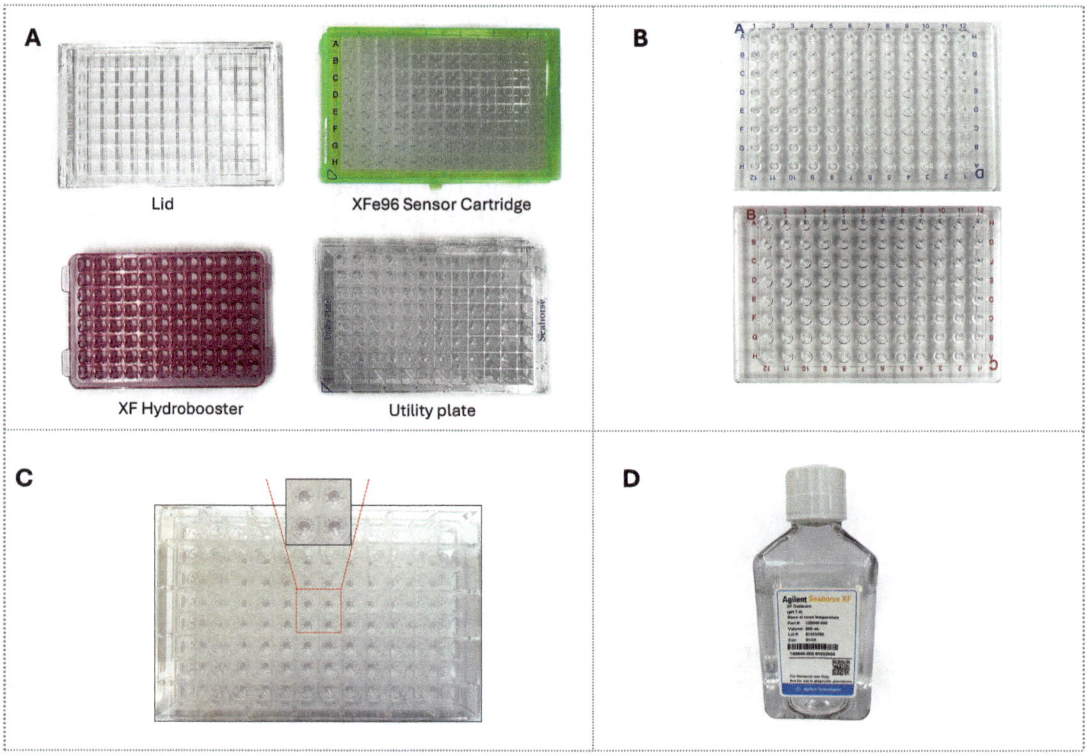

Fig. 1 Key components for Seahorse analysis: Agilent Seahorse XFe96 Spheroid FluxPaks components **a**. Agilent Port guides for loading the ports **b**. Agilent Seahorse XFe96 Spheroid Microplates **c**. Agilent Seahorse XF calibrant **d**

2.1 Equipment and Materials for Bioenergetic Analysis

Analyzer. This microplate's distinctive design empowers precise positioning of individual zebrafish larvae, streamlining the execution of gold standard metabolic assessments. This ground-breaking product offers a unique capability for conducting highly sensitive, real-time metabolic measurements in individual zebrafish larvae, providing a dependable approach for evaluating metabolic functionality.

1. Agilent Seahorse XFe96 Analyzer.
2. Non-CO_2 incubator.
3. Dissecting or inverted microscope (for verifying the correct positioning of zebrafish).
4. Agilent Seahorse XF Base Medium (without phenol red) (Agilent #103335-100). Store at 4 °C and protected from light.
5. Agilent Seahorse XFe96 Spheroid FluxPak (Agilent #102905-100) (Fig. 1), composed of Agilent Seahorse XFe96 Flux assay kit, containing Cartridge lid, XFe96 Sensor Cartridge, XF Hydrobooster and Utility plate (Fig. 1a), Agilent Port Guides (Fig. 1b), Agilent Seahorse XFe96 Spheroid Microplates

(Fig. 1c), and Agilent Seahorse XF calibrant (500 mL) (Fig. 1d).

2.2 Mitochondrial Inhibitor Stock Preparation

2.2.1 Oligomycin Stock Solution

1. Dissolve 5 mg Oligomycin powder (Sigma, O4876) in 2.53 mL of DMSO to make a 2.5 mM solution.
2. Aliquot 250 μL per tube and store at −80 °C for up to 12 months (do not freeze-thaw).
3. Oligomycin will be used in Seahorse injection port A to inhibit ATP synthase.

2.2.2 FCCP Stock Solution

1. Dissolve 10 mg of FCCP powder (Sigma, C2920) in 1.57 mL of DMSO to make a 25 mM solution.
2. Further dilute 25 mM FCCP to 2.5 mM with DMSO, by adding 100 μL of FCCP (25 mM) to 900 μL of DMSO.
3. Aliquot 100 μL of the 2.5 mM FCCP stock per tube and store at −80 °C for up to 6 months (do not freeze-thaw).
4. FCCP will be used in Seahorse injection port B to collapse the proton gradient.

2.2.3 Rotenone and Antimycin A Stock Solutions

1. Dissolve 1 g of Rotenone powder (Sigma, R8875) in 25.35 mL of DMSO to make a 100 mM solution.
2. Further dilute 100 mM Rotenone to 25 mM with DMSO, by adding 1.25 mL of Rotenone (100 mM) to 3.75 mL of DMSO. Aliquot into 10 μL per tube and store at −80 °C for up to 12 months (do not freeze-thaw).
3. Dissolve 25 mg of Antimycin A powder (Sigma, A8674) with 4.7 mL of DMSO to make a 10 mM solution. Further dilute 10 mM Antimycin A to 1.5 mM with DMSO, by adding 150 μL of Antimycin A to 850 μL of DMSO.
4. Aliquot into 30 μL per tube and store at −80 °C for up to 12 months (do not freeze-thaw).
5. Rotenone/Antimycin will be used together in Seahorse injection port C to block mitochondrial respiration. *See* **Note 1** for information on using Agilent compounds.

2.3 Equipment and Materials for Zebrafish Preparation

2.3.1 E3 Media Preparation

1. To prepare a 60× stock solution of E3 media, dissolve the compounds (34.8 g NaCl, 1.6 g KCl, 5.8 g $CaCl_2{\cdot}2H_2O$, 9.78 g $MgCl_2{\cdot}6H_2O$) in H_2O, to a final volume of 2 L.
2. Adjust the pH to 7.2 with NaOH.
3. Autoclave.
4. To prepare 1× medium, dilute 16.67 mL of the 60× stock to 1 L.
5. Add 100 μL of 1% methylene blue.

6. E3 media will be used to dilute the Tricaine for anesthetizing, and Pronase for dechorionating, the zebrafish.

2.3.2 Tricaine Preparation

1. Dissolve 4 g of Tricaine powder (a.k.a. MS-222) in 20.8 mL of Trizma Base and 0.8 L of distilled water.
2. The pH of Tricaine Solution should be 7.0–7.2. Bring the volume up to 1 L to make a 0.4% stock solution.
3. Store at 4 °C for up to 6 months [22].
4. Tricaine will be used to anesthetize the zebrafish by diluting the 0.4% stock 1 in 10 in E3 media to a working concentration of 0.04%.

2.3.3 Pronase Preparation

1. To make 0.5 mg/mL Pronase working concentration, mix 125 μL of 20 mg/mL Pronase stock with 5 mL of fish water or E3 with methylene blue in a 94 × 16 mm Petri dish.
2. 0.5 mg/mL Pronase can be used to dechorionate the zebrafish.

3 Methods

3.1 Preparation on the Day Before the Assay

3.1.1 Cartridge Preparation (Hydration)

The Seahorse cartridges must be hydrated (Fig. 2). While carrying out these steps, ensure the XFe96 DM Cartridge is put upside down (sensors facing up as shown in Fig. 2a and b) to prevent damaging it:

1. Remove the lid (clear), XFe96 Sensor Cartridge (green), and XF Hydrobooster (pink) from the Utility Plate (clear).
2. Add 200 μL of XF calibrant solution to the wells of the Utility Plate (clear). Use of a multichannel pipette is recommended.
3. Place the XF Hydrobooster (pink) back onto the Utility Plate (clear) and ensure a secure fit by pressing it down firmly.
4. Replace the XFe96 Sensor Cartridge (green) onto the Utility Plate (clear) through the openings in the XF Hydrobooster (pink), ensuring the sensors are fully submerged in the XF calibrant.
5. Confirm that the XF calibrant level is sufficient to ensure the sensors remain submerged.
6. Place the lid back on the fully assembled plates.
7. Transfer the setup to a non-CO_2 humidified incubator set at 37 °C and let it stand overnight (*see* **Note 2**). To prevent XF calibrant evaporation, ensure that the incubator is adequately humidified overnight.
8. Set the Seahorse machine to 28 °C in preparation for the next day (*see* **Note 3**).

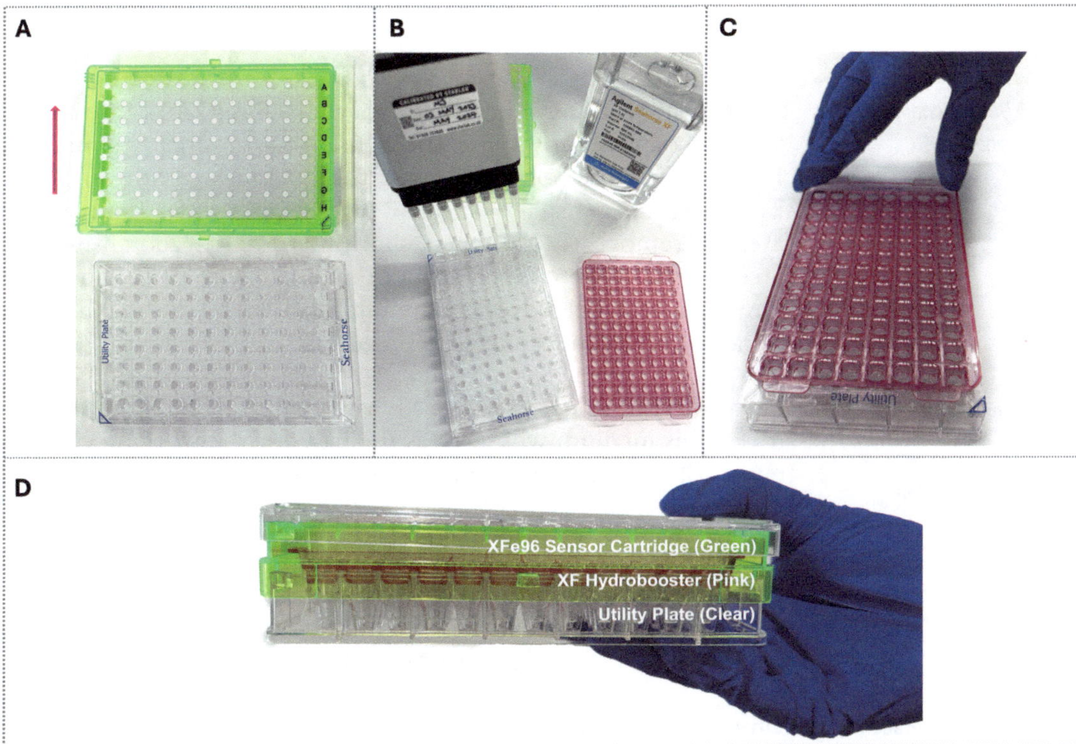

Fig. 2 Cartridge rehydration procedure: XF Hydrobooster (pink) and XFe96 Sensor Cartridge (green) are removed from the Utility Plate. The XFe96 Sensor Cartridge (green) is placed with sensors facing upward to prevent damage **a**. XF Calibrant is added to the designated area on the Utility plate (clear) using a multichannel pipette **b**. The XF Hydrobooster (pink) is securely reattached to the Utility Plate (clear) **c**. The XFe96 Sensor Cartridge (green) is carefully placed back onto the XF Hydrobooster (pink)/Utility Plate (clear), ensuring proper alignment. The assembly is covered with the lid (clear) to complete the preparation process **d**

3.2 Zebrafish Dechorionation

The conventional techniques for chorion removal include a manual extraction using forceps, and an alternative enzymatic method involving Pronase treatment. Depending on the scale of the experiment, the manual removal approach can be quite demanding and time-consuming due to the precision and delicacy required. A potentially quicker method is the application of Pronase, which uses a mixture of proteolytic enzymes derived from *Streptomyces griseus* to enzymatically soften the chorion until it reaches a point of rupture, consequently allowing the zebrafish to be released. Both techniques are suitable for use on 2 dpf zebrafish.

3.2.1 Manual Dechorionation

Raise larvae in system aquaria water or E3 with 1% methylene blue. When opting for manual chorion removal, forceps are employed to gently pinch the chorion's surface, and a careful outward pull is applied to create a rupture, then one forceps is moved so that a second perpendicular rupture is made, thereby freeing the zebrafish.

3.2.2 Dechorionation Using Pronase

Transfer the working solution of Pronase (0.5 mg/mL) into a small Petri dish and gently place the zebrafish, still enclosed in their chorion, into the solution. Swirl the zebrafish gently until the chorion softens (usually within 3–5 min at 2 dpf). If necessary, carefully remove the softened chorion from the zebrafish using forceps under a stereomicroscope. Immediately transfer the zebrafish into a new Petri dish containing fresh E3 medium. Rinse the zebrafish three times in fresh E3 medium.

3.3 The Day of the Assay

3.3.1 Mitochondrial Inhibitor Preparation

Prepare 40 mL using 36 mL of XF base medium, plus 4 mL tricaine stock. This medium will serve as the base for the working concentration of mitochondrial inhibitors and for loading the spheroid plate wells.

Port A: Oligomycin Preparation

Dilute the 2.5 mM Oligomycin stock solution by adding 200 μL to 1800 μL XF base medium containing Tricaine, according to the concentrations in Table 1. *See* **Note 4** for calculations.

Port B: FCCP Preparation

Dilute the 2.5 mM FCCP stock solution by adding 112.5 μL to 3187.5 μL of XF base medium containing Tricaine, according to the concentrations in Table 1. *See* **Note 4** for calculations.

Port C: Rotenone and Antimycin A Preparation

Dilute the 25 mM Rotenone and 1.5 mM Antimycin A stock solutions by adding 2.25 μL of Rotenone and 37.5 μL of Antimycin A stock solutions to 3710.25 μL XF base medium containing Tricaine, according to the concentrations in Table 1. *See* **Note 4** for calculations.

3.3.2 XFe96 Sensor Cartridge Preparation for Calibration

1. Retrieve the fully assembled plates from the incubator (as prepared in Subheading 3.1 and Fig. 2).

Table 1
Compound volumes to add to each port within each well

Port	Compound	Stock solution (C_1)	Assay medium volume (V_2)	Working concentration (C_2)	Port volume
A	Oligomycin	2.5 mM	200 μL	25 μM	20 μL
B	FCCP	2.5 mM	222 μL	8.5 μM	22 μL
C	Rotenone/Antimycin A	25 mM/1.5 mM	247 μL	1.5 μM	25 μL

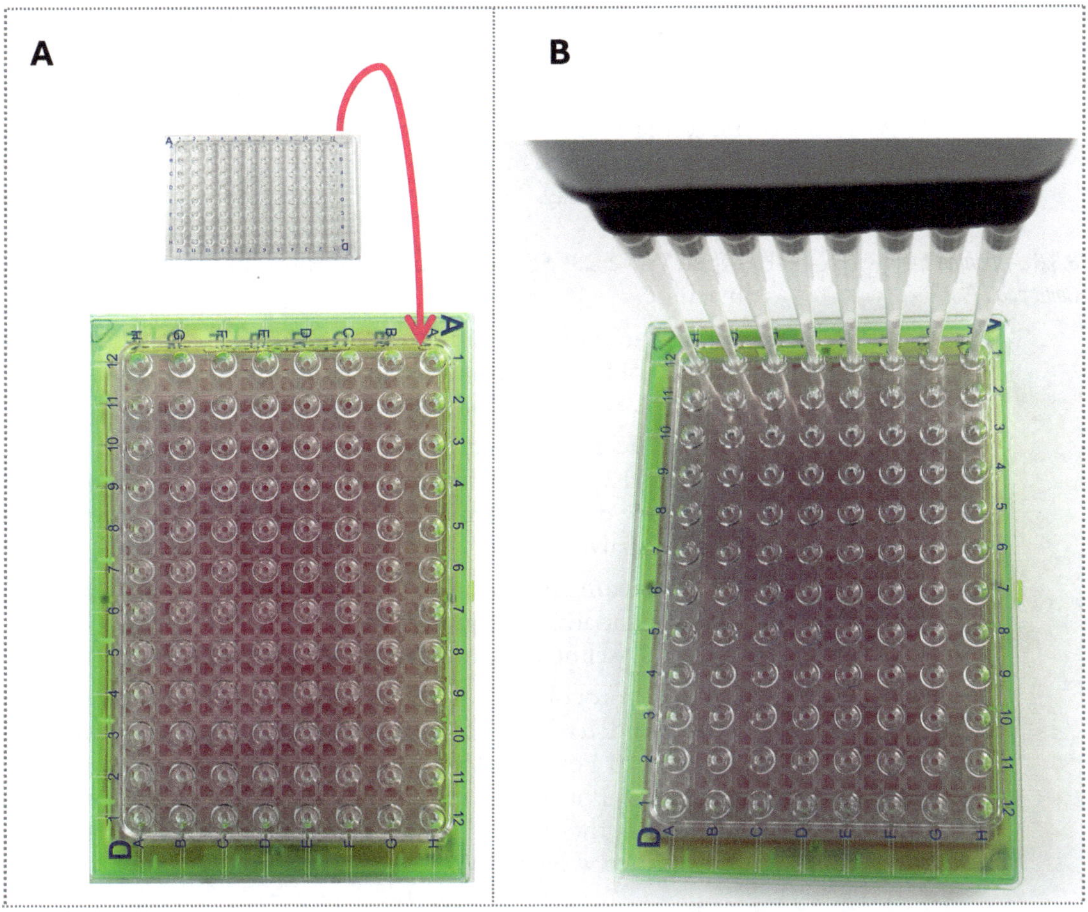

Fig. 3 Drug loading procedure: The alignment of guide holes with Port A enables filling via a multichannel pipette. To use the guides to load Port A, the guide is positioned on the plate with the "A" marking situated in the top left-hand corner. For Port D, rotate the guide 180 degrees, positioning the "D" marking in the top left-hand corner to align with Port D. Similar procedures apply for ports B and C using separate guides **a**. Once the port guide has been applied, use a multichannel pipette to load the drug into that port. Load the other ports in the same way, rotating the guides as appropriate **b**.

2. Take off the XFe96 Sensor Cartridge (green) (ensure the XFe96 Sensor Cartridge is placed upside down to prevent damaging the sensors).
3. Gently lift and detach the XF Hydrobooster (pink) from one corner of the plate. This can be disposed of.
4. Reattach the XFe96 Sensor Cartridge (green) to the Utility Plate (clear).
5. Use the Port Guides provided by Agilent and carefully pipette compounds, one compound at a time, using a multichannel pipette into ports A–C of each well of the XFe96 Sensor Cartridge (green), making sure not to insert the pipette fully into the ports (the solution stays in the ports by surface

tension) (Fig. 3). The volumes required for each port of each well are shown in Table 1. Fill all the ports with all the compounds, including the ports for the wells you are not using.

6. Place the loaded XFe96 Sensor Cartridge (green) and Utility Plate (clear) back into the 37 °C incubator for at least 30 min prior to the assay. This plate will be used in the calibration step.

3.3.3 Running the Calibration Plate

1. Open Wave 2.4 Software on the monitor linked to the Seahorse machine.
2. Set up the protocol using the "Mito Stress test Seahorse template" (blue) or modify/load your assay protocol (green) (Fig. 4a), so that the protocol matches that in Table 2.
3. Define the assay layout in the Plate Map tab (Fig. 4b), where A1 and A12, H1 and H12 correspond to the background wells. They have default Plate Map coordinates based on the type of XFe96 Analyzer.
4. Each Instrument Protocol comprises the following steps: Calibrate (mandatory), Equilibrate (strongly recommended for all assays), Baseline Measurement Cycle (Fig. 4c).
5. On the Protocol tab, edit the measurement cycle commands to correspond to Table 2 (Fig. 5a). Ensure that Oligomycin is assigned to Port A, FCCP to Port B and Rotenone and Antimycin A to Port C. Measurement Cycle Commands: Mix: Raise and lower the XFe96 Sensor Cartridge to mix analytes and reintroduce oxygen; Wait: Delay period following the Mix

Fig. 4 Workflow steps for Mito Stress Test assay analysis: This figure illustrates the sequential procedures involved in conducting and analyzing the Mito Stress Test assay using the Wave 2.4 Software and Agilent Seahorse Analytics software. Select the template for the Mito Stress Test assay between default templates (in blue) and previously modified templates (in green) within the Wave 2.4 Software interface **a**. The plate layout for the Mito Stress Test assay is defined, specifying the arrangement of samples and controls within the assay plate **b**. The series of instrument protocol steps required to execute the Mito Stress Test assay are outlined, detailing the parameters and actions to be performed during the experimental procedure **c**

Table 2
Mito stress protocol parameters

Stage	Cycles	Mixing time (min)	Waiting time (min)	Measuring time (min:s)
Basal respiration	10	2	0	2:30
Oligomycin	15	2	0	3:00
FCCP	4	3	0	3:00
Rotenone/Antimycin A	5	2	0	3:00

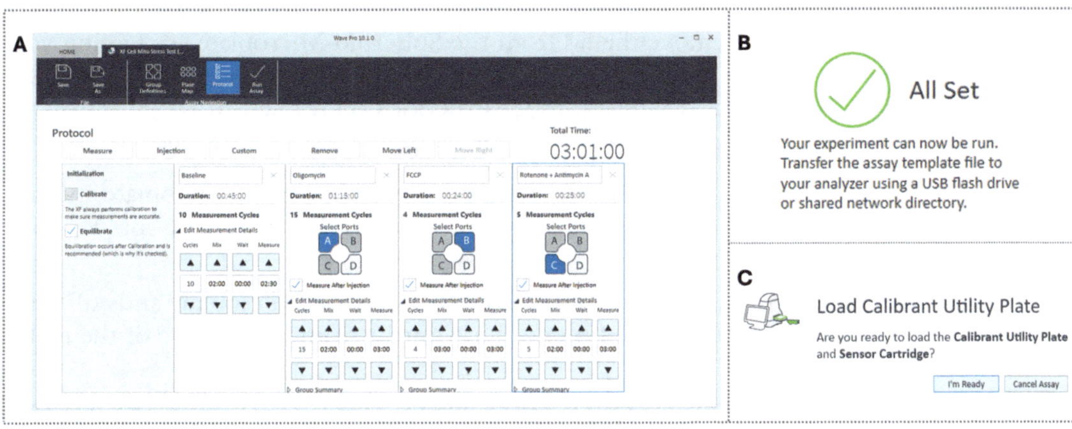

Fig. 5 Software interface for preparing assay execution: Users are shown modifying measurement cycle commands within the software interface to tailor assay parameters according to experimental requirements **a**. Indication that assay is ready to be run **b**. The software prompts asking if you are ready to load the utility plate **c**

step; Measure: Record analyte flux in the transient microchamber (Table 2, Fig. 5a).

6. When the protocol is finalized and saved, Click Start on the Run Assay tab (Fig. 5b).
7. Carefully transfer the XFe96 Sensor Cartridge (green) together with the Utility Plate (clear) and load in the proper orientation. The XFe96 Analyzer reads and records the Cell Plate and XFe96 Sensor Cartridge barcodes before beginning an assay.
8. Remove the lid and loading guide before loading the plate into Analyzer (Fig. 5c) and click "I am ready"; the calibration will start.
9. When the calibration is completed, replace the Utility Plate with the Spheroid Microplate containing the embryos/larvae (*see* Subheading 3.3.4 for preparation) according to the software instructions. The XFe96 Sensor Cartridge will be retained in the machine.

3.3.4 Zebrafish Sedation and Transfer to the Seahorse Spheroid Microplate

1. Incubate larvae in a Petri dish containing E3 containing Tricaine for 5–10 min.
2. Add 170 μL of XF base medium containing Tricaine to each well of the Spheroid Microplate (Fig. 1c), including the ones you are not planning to use (you can do this while the plate is calibrating; *see* above).
3. Transfer the zebrafish (one per well) with a maximum of 10 μL of Tricaine working solution into the Spheroid Microplates with a 1000 μL pipette tip with the end cut off (*see* **Note 5**).
4. Put the lid back on the Spheroid Microplate.

3.3.5 Running the Spheroid Microplate Containing the Zebrafish

1. Remove the lid from the Spheroid Microplate containing your zebrafish and place it into the machine where the utility plate was, following the instructions on the screen.
2. Start the Mito Stress Test run.
3. At the end of the run, remove the Spheroid Microplate and save the data (this can be saved as Excel, Prism, or Seahorse Analytics).
4. Remove the larvae from the Spheroid Microplate and cull using an approved method in your institution. Dispose of the plate.

3.4 Data Analysis

1. Agilent Seahorse Analytics provide their own Analytic software. Make an account with Agilent Seahorse Analytics and log in (https://www.agilent.com/account/login).
2. To open your saved data file, go to the home tab and upload your file (Fig. 6a).
3. Select the widgets under the "Assay Kit Companion View: XF Cell Mito Stress Test" that are relevant to your interests (Fig. 6b) and click "Add view."
4. To remove outliers from your data, edit the mitochondrial respiration widget, as it has access to the plate map.
5. Using the flag tool (each tool is labeled with its name on the software as you hover over it), identify outliers or empty wells with a flag. Then click the "hide flag well" tool to remove these from the analysis (Fig. 6c).
6. When you return to the XF Cell Mito Stress Test View, the graphs on the remaining widgets will correspond to the edited plate map where you have removed outlier(s).
7. By clicking on any widget, you can find the mean and standard deviations for each of the groups.
8. These graphs will allow you to compare your groups for the different parameters, e.g., basal respiration.
9. Export the data from the analytic software (website) (Fig. 6d) and transfer them into Graph Pad Prism (or similar) for statistical analysis (Fig. 7, *see* **Note 6**).

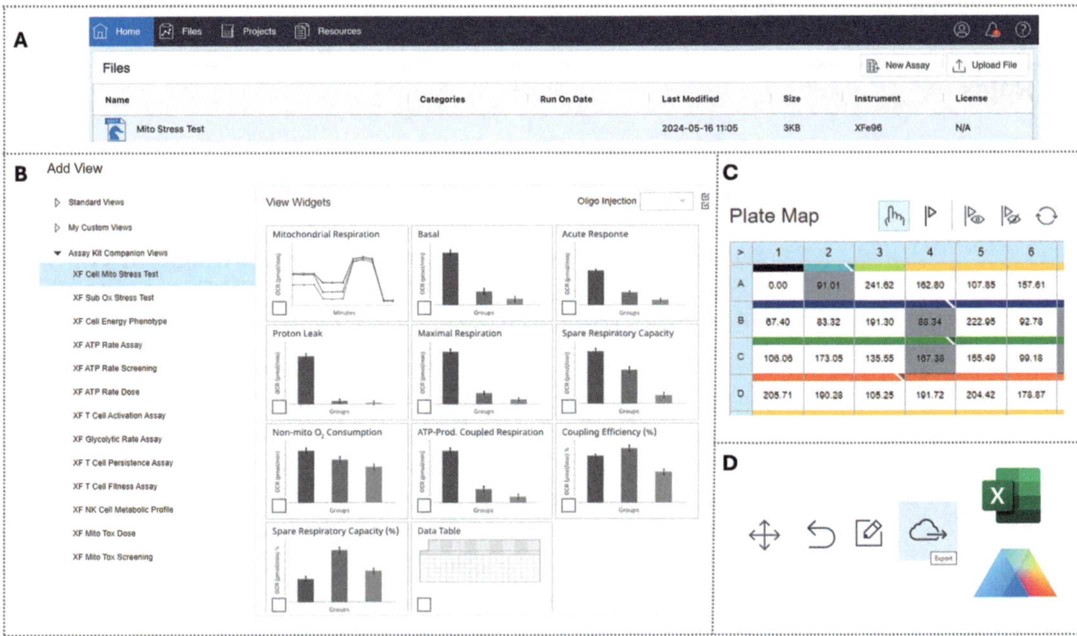

Fig. 6 Using Agilent Seahorse Analytics software online: Open an assay file on the Agilent Seahorse Analytics software **a**. Visual example of the analytic software "Widget Page" **b**. Removing empty wells or outliers from the analysis using the flag tool **c**. Exporting the data from the analytic software **d**

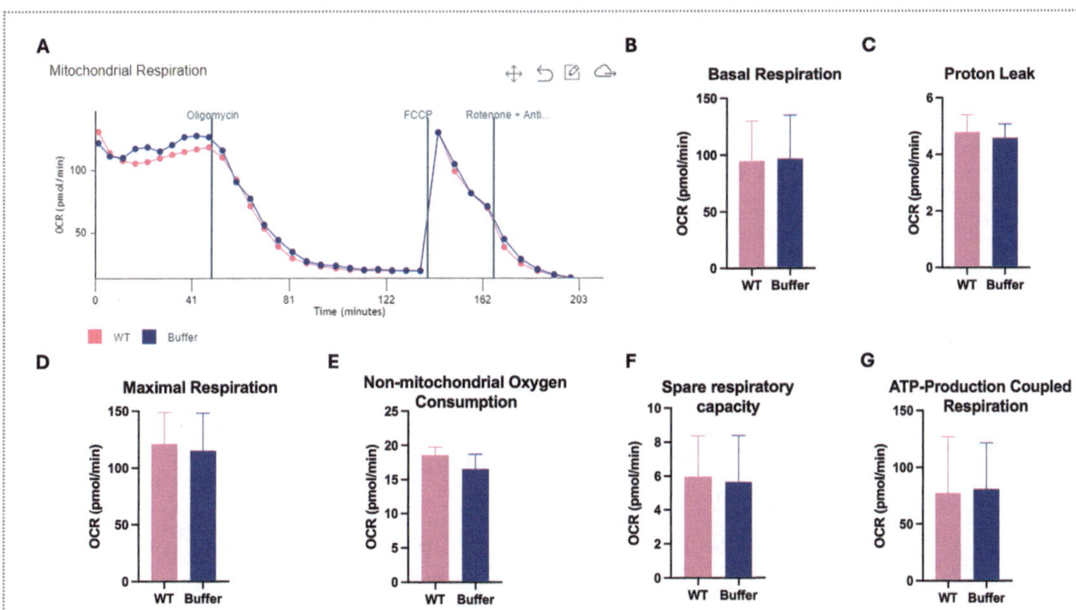

Fig. 7 Analysis of mitochondrial respiration in wild-type (WT) zebrafish larvae at 2 dpf using Seahorse XF Mito Stress Test. Oxygen Consumption Rate (OCR) profiles for WT and buffer-injected WT zebrafish larvae at 2 dpf were assessed using the Seahorse XF Mito Stress Test. The assay evaluated several parameters of mitochondrial function, including the kinetics **a**, basal respiration **b**, proton leak **c**, maximal respiration **d**, non-mitochondrial respiration **e**, spare respiratory capacity **f**, and ATP production **g**. Data analysis was conducted using Agilent Seahorse analytic software and GraphPad Prism software **b**–**g**

4 Notes

1. If you use the Agilent Mito stress Kit, see their instructions, and you can prepare the mitochondrial inhibitors on the day of the assay (Seahorse XFp Cell Mito Stress TestMito Stress test Kit User Guide (agilent.com) https://www.agilent.com/cs/library/usermanuals/public/XFp_Cell_Mito_Stress_Test_Kit_User_Guide.pdf).
2. Manufacturer's guidance recommends soaking the XFe96 Sensor Cartridge for an entire night at 37 °C to achieve the best results. While it is possible to hydrate the cartridge for a minimum of 4 h at 37 °C, it is worth noting that doing so may compromise the quality of the data. On the other hand, the sensor can be hydrated for up to 72 h, but extended hydration might necessitate additional calibrant solution to compensate for potential evaporation.
3. By default, the machine temperature is set to 37 °C, which works best for cells. Changing the target temperature requires overnight equilibration to the new set point. The machine is sensitive to room temperature fluctuations, so it is essential to maintain a constant room temperature.
4. To prepare the inhibitors, we used the following calculations:

Oligomycin:

Stock solution concentration (C1): 2.5 mM; Final concentration (C2): 25 μM; Port A Volume: 20 μL.

$C1V1 = C2V2$

$C1 = 2500\ \mu M$

$V1 = \times\ \mu L$

$C2 = 25\ \mu M$

$V2 = 200\ \mu L$

$\times = 25 \times 200/2500\ \mu L = 2\ \mu L$

$2\ \mu L \times 100 = 200\ \mu L$ (to prepare sufficient solution for a 96-well plate, including excess)

FCCP:

Stock solution concentration: 2.5 mM; final concentration: 8.5 μM; Port B Volume: 22 μL.

$C1V1 = C2V2$

$C1 = 2500\ \mu M$

$V1 = \times\ \mu L$

$C2 = 8.5\ \mu M$

$V2 = 222\ \mu L$

$\times = 8.5 \times 222/2500\ \mu L = 0.75\ \mu L$

$0.75\ \mu L \times 100 = 75\ \mu L$ (to prepare sufficient solution for a 96-well plate, including excess)

Rotenone/Antimycin A

Stock solution concentrations: Rotenone: 25 mM; Antimycin A: 1.5 mM; Final concentration: 1.5 μM for both Rotenone and Antimycin separately; Port C Volume: 25 μL.

Rotenone

$C1V1 = C2V2$

$C1 = 25{,}000\ \mu M$

$V1 = \times\ \mu L$

$C2 = 1.5\ \mu M$

$V2 = 247\ \mu L$

$\times = 1.5 \times 247\ /\ 25{,}000\ \mu L = 0.015\ \mu L$

$0.015\ \mu L \times 100 = 1.5\ \mu L$ (to prepare sufficient solution for a 96-well plate, including excess)

Antimycin A

$C1V1 = C2V2$

$C1 = 1500\ \mu M$

$V1 = \times\ \mu L$

$C2 = 1.5\ \mu M$

$V2 = 247\ \mu L$

$\times = 1.5 \times 247/1500\ \mu L = 0.25\ \mu L$

$0.25\ \mu L \times 100 = 25\ \mu L$ (to prepare sufficient solution for a 96-well plate, including excess)

5. We have observed that some wells, particularly those situated at the outer edge of the plate, exhibit diminished functionality, leading to inconsistencies in data quality. This phenomenon likely arises from factors such as temperature gradients, media evaporation, and edge effects. Addressing these challenges through protocol adjustments and optimization strategies is essential for improving the reliability and reproducibility of metabolic measurements obtained from outer edge wells. In addition to this, due to plate-to-plate variation, ensure that all your experimental conditions are represented on one plate.
6. Seahorse Agilent analytic software algorithm is set to average all the measured values. However, due to the nature of the curve produced after FCCP addition to 2 dpf zebrafish, only the highest peak for maximal respiration should be used for the analysis.

Acknowledgments

We thank Claire Thornton and Imelda McGonnell (both from the Royal Veterinary College, London, UK) for advice and co-authoring grant applications with CR, leading to an MRC grant (MC_PC_MR/X01214X/1) for equipment purchase and a PhD studentship from the Anatomical Society for EY. This work was also supported by an MRC grant (MR/V033956/1) to CR for salary support for VM. The grant is led by Sara Mole (UCL, London, UK).

References

1. Palmer DN, Barry LA, Tyynela J, Cooper JD (2013) NCL disease mechanisms. Biochim Biophys Acta 1832(11):1882–1893. https://doi.org/10.1016/j.bbadis.2013.05.014
2. Johnson TB, Cain JT, White KA et al (2019) Therapeutic landscape for Batten disease: current treatments and future prospects. Nat Rev Neurol 15(3):161–178. https://doi.org/10.1038/s41582-019-0138-8
3. Kousi M, Lehesjoki AE, Mole SE (2012) Update of the mutation spectrum and clinical correlations of over 360 mutations in eight genes that underlie the neuronal ceroid lipofuscinoses. Hum Mutat 33(1):42–63. https://doi.org/10.1002/humu.21624
4. Nickel M, Simonati A, Jacoby D et al (2018) Disease characteristics and progression in patients with late-infantile neuronal ceroid lipofuscinosis type 2 (CLN2) disease: an observational cohort study. Lancet Child Adolesc Health 2(8):582–590. https://doi.org/10.1016/S2352-4642(18)30179-2
5. Jolly RD, Brown S, Das AM, Walkley SU (2002) Mitochondrial dysfunction in the neuronal ceroid-lipofuscinoses (Batten disease). Neurochem Int 40(6):565–571. https://doi.org/10.1016/s0197-0186(01)00128-0
6. Onyenwoke RU, Brenman JE (2015) Lysosomal storage diseases-regulating neurodegeneration. J Exp Neurosci 9(Suppl 2):81–91. https://doi.org/10.4137/JEN.S25475
7. Gu X, Ma Y, Liu Y, Wan Q (2021) Measurement of mitochondrial respiration in adherent cells by Seahorse XF96 Cell Mito Stress Test. STAR Protoc 2(1):100245. https://doi.org/10.1016/j.xpro.2020.100245
8. Saffari A, Kolker S, Hoffmann GF, Ebrahimi-Fakhari D (2017) Linking mitochondrial dysfunction to neurodegeneration in lysosomal storage diseases. J Inherit Metab Dis 40(5):631–640. https://doi.org/10.1007/s10545-017-0048-0
9. Scifo E, Szwajda A, Soliymani R et al (2015) Proteomic analysis of the palmitoyl protein thioesterase 1 interactome in SH-SY5Y human neuroblastoma cells. J Proteome 123:42–53. https://doi.org/10.1016/j.jprot.2015.03.038
10. Tikka S, Monogioudi E, Gotsopoulos A et al (2016) Proteomic profiling in the brain of CLN1 disease model reveals affected functional modules. NeuroMolecular Med 18(1):109–133. https://doi.org/10.1007/s12017-015-8382-6
11. Doccini S, Morani F, Nesti C et al (2020) Proteomic and functional analyses in disease models reveal CLN5 protein involvement in mitochondrial dysfunction. Cell Death Discov 6:18. https://doi.org/10.1038/s41420-020-0250-y
12. Lopez-Fabuel I, Garcia-Macia M, Buondelmonte C et al (2022) Aberrant upregulation of the glycolytic enzyme PFKFB3 in CLN7 neuronal ceroid lipofuscinosis. Nat Commun 13(1):536. https://doi.org/10.1038/s41467-022-28191-1
13. Pezzini F, Bianchi M, Benfatto S et al (2017) The networks of genes encoding palmitoylated proteins in axonal and synaptic compartments are affected in PPT1 overexpressing neuronal-like cells. Front Mol Neurosci 10:266. https://doi.org/10.3389/fnmol.2017.00266
14. Pezzini F, Gismondi F, Tessa A et al (2011) Involvement of the mitochondrial compartment in human NCL fibroblasts. Biochem Biophys Res Commun 416(1–2):159–164. https://doi.org/10.1016/j.bbrc.2011.11.016
15. Bosch ME, Kielian T (2019) Astrocytes in juvenile neuronal ceroid lipofuscinosis (CLN3) display metabolic and calcium signaling

abnormalities. J Neurochem 148(5):612–624. https://doi.org/10.1111/jnc.14545

16. Desousa BR, Kim KK, Jones AE et al (2023) Calculation of ATP production rates using the Seahorse XF Analyzer. EMBO Rep 24(10): e56380. https://doi.org/10.15252/embr.202256380
17. Mahmood F, Fu S, Cooke J et al (2013) A zebrafish model of CLN2 disease is deficient in tripeptidyl peptidase 1 and displays progressive neurodegeneration accompanied by a reduction in proliferation. Brain 136(Pt 5): 1488–1507. https://doi.org/10.1093/brain/awt043
18. Wager K, Zdebik AA, Fu S et al (2016) Neurodegeneration and epilepsy in a zebrafish model of CLN3 disease (Batten disease). PLoS One 11(6):e0157365. https://doi.org/10.1371/journal.pone.0157365
19. Crouzier L, Denus M, Richard EM et al (2021) Sigma-1 receptor is critical for mitochondrial activity and unfolded protein response in larval zebrafish. Int J Mol Sci 22(20). https://doi.org/10.3390/ijms222011049
20. Kumar A, Singh N, Goswami M et al (2016) Establishment and characterization of a new muscle cell line of zebrafish (Danio rerio) as an in vitro model for gene expression studies. Anim Biotechnol 27(3):166–173. https://doi.org/10.1080/10495398.2016.1147455
21. Rollwitz E, Jastroch M (2021) Plate-based respirometry to assess thermal sensitivity of zebrafish embryo bioenergetics in situ. Front Physiol 12:746367. https://doi.org/10.3389/fphys.2021.746367
22. Katz EM, Chu DK, Casey KM et al (2020) The stability and efficacy of tricaine methanesulfonate (MS222) solution after long-term storage. J Am Assoc Lab Anim Sci 59(4):393–400. https://doi.org/10.30802/AALAS-JAALAS-19-000067

Chapter 16

Immunostaining Methods for Lysosomal Storage Disorder Research

Ewa A. Ziółkowska, Sophie H. Wang, Hemanth R. Nelvagal, and Jonathan D. Cooper

Abstract

Immunostaining enables the detection of proteins and antigens in histological preparations due to specific antigen–antibody interactions. Such visualization can be performed by immunoperoxidase methods, but more recently, we have switched to immunofluorescence methods that offer superior sensitivity and simultaneous detection of multiple antigens. Key to this approach is a counterstaining method that masks tissue histofluorescence and the autofluorescent storage material that accumulates in the disorders we study. Such methods allow the analysis of the structural organization of lysosomes and the impact of their dysfunction on cells and tissues. In this chapter, we present an optimized staining protocol that can serve as a standard procedure in histological examinations, ensuring high quality and consistency of quantitative results.

Key words Immunohistochemistry, Immunofluorescence, Histology, Primary antibody, Secondary antibody, Staining, TrueBlack®

1 Introduction

Immunohistochemistry (IHC) is one of the fundamental tissue analysis techniques used in both biological research and pathological diagnosis. The main goal is to determine the distribution and abundance of cellular components in tissue sections via the visualization of proteins and other antigens [1–4]. This method relies largely on the specificity of antigen–antibody interactions, which are then amplified via a species-specific secondary antibody and visualized using a colored enzyme reaction product (most typically immunoperoxidase methods using diaminobenzidine dihydrochloride) or a fluorescently labeled secondary antibody or streptavidin [5–7]. The foundations for this theory were developed in 1941 by Albert Coons et al. [8]. However, significant refinements to these techniques for fixed tissues took place in the 1970s and

Stephanie M. Hughes and Indranil Basak (eds.), *Lysosomes: Methods and Protocols*, Methods in Molecular Biology, vol. 2976, https://doi.org/10.1007/978-1-0716-4844-5_16,

1980s [9–13]. Key principles involve adequate fixation of samples to preserve morphology, while still retaining antigenicity. Also important is the prevention, or at least minimizing, of unwanted nonspecific background staining that may be the result of nonspecific binding of either the primary or secondary antibodies to tissue components. This is typically performed by blocking such immunoglobulin-antigen interactions by saturating nonspecific binding sites by high concentrations of normal serum, but protein-antigen interactions may also be blocked by bovine serum albumin or milk powder [1, 2, 4, 5]. A key problem we faced in analyzing LSD tissue is the pronounced autofluorescence of the storage material that accumulates in lysosomes, most notably in the neuronal ceroid lipofuscinoses or Batten disease. Because this material autofluoresces across all commonly used filter sets and laser lines, this typically obscures any immunofluorescence (IF) staining. However, we now routinely use the counterstain TrueBlack®, which quenches this autofluorescence and also blocks the histofluorescence evident in aldehyde-fixed tissues. This has revolutionized our ability to routinely and reliably stain sections for quantitative analyses, most typically via thresholding image analysis or stereology.

In relation to LSDs, we can routinely stain components of the lysosome itself, the material that accumulates within the lysosome, or therapeutically delivered lysosomal enzymes that are deficient in these disorders [14, 15]. In addition, we can stain for immunohistochemical markers of either different neuron subtypes that die in these disorders, or for the associated reactive changes that occur in glial cell populations, most often being markers of astrocytosis or microglial activation. Monitoring how any of these markers change in different central nervous system (CNS) regions during disease progression provides baseline data against which the efficacy of different experimental therapies can be judged. However, it is crucial to maintain an optimized and repeatable staining scheme, which should use high-quality, specifically selected reagents. A representative suggested method for slide-mounted immunofluorescence is presented below.

2 Materials

Prepare all reagents using distilled water and analytical grade reagents. Prepare all reagents at room temperature and store at 4 °C (unless indicated otherwise).

2.1 Reagents and Materials

1. Tris-buffered saline (TBS, 1×).
2. 4% TBS-T: Mix 50 mL of TBS 10× with 450 mL of dH_2O. Remove 20 mL of TBS 1× and add 20 g of Triton

X-100. Mix for 30 min, then add TBS 1× until total volume is 500 mL, pH 7.6 (*see* **Note 1**).

3. Blocking buffer: 15% Normal Goat Serum (NGS) in 4% TBS-T. For 500 μL of buffer, mix 75 μL of NGS and 425 μL of 4% TBS-T (*see* **Note 2**).
4. Incubation buffer: 10% NGS in 4% TBS-T. For 500 μL of buffer, mix 50 μL of NGS and 450 μL of 4% TBS-T (*see* **Note 3**).
5. TrueBlack® counterstain: For 500 μL of buffer, add 475 μL of 70% EtOH and 25 μL of TrueBlack® (*see* **Note 4**).
6. Light-tight plastic chamber for staining of slides (*see* **Note 5**).
7. Hydrophobic Barrier PAP Pen.
8. Fluoromount-G mounting medium containing DAPI (4′,6-diamidino-2-phenylindole).

3 Methods

The staining technique discussed applies to slide-mounted sections mounted on ColorFrost Plus microscope slides. It is assumed that approximately 500 μL of reagent is needed for one slide; depending on the number of slides, this volume should be multiplied accordingly. At no stage should the preparations be allowed to dry. All staining takes place at room temperature. Most important is to use a consistent time to allow slide-mounted sections to dry on the microscope slide, with 30 min recommended, to prevent background fluorescence due to sections becoming too dry. At the end of this period, slides should be kept covered in a slide rack in Tris-buffered saline at 4 °C, prior to immunostaining. In our experience, provided the appropriate ColorFrost Plus slides are used, slide-mounted sections can be kept in this fashion for several weeks, if not months, before staining. This allows mounting of sections to be done over several days and then stained in appropriately sized batches to minimize variation.

3.1 Marking the Staining Area

To facilitate the process of applying reagents to sections, mark the area around them with a hydrophobic Barrier PAP Pen. Slides should be placed in a staining container (*see* **Note 6**).

3.2 Blocking Incubation

1. 15% NGS in 4% TBS-T. Prepare the buffer as described above in an amount corresponding to the number of slides. Apply it to the slides using 500 μL/slide, cover the staining chamber again, and leave for 1 h (*see* **Note 7**).
2. After incubating the sections, remove the blocking buffer by pouring it onto a tissue (*see* **Note 8**).

3.3 Primary Incubation

1. Prepare 10% NGS in 4% TBS-T according to the recipe above using 500 μL/slide. Then, add the selected primary antibody, depending on the protein being analyzed, to the prepared buffer. For example, to analyze storage material in brain tissue or other tissues, you can use the rabbit polyclonal antibody to ATP synthase subunit C (*see* **Note 9**).
2. Then, apply the prepared buffer to the slides, cover, and set aside for 2 h (*see* **Note 10**).
3. After this time, remove the primary antibody, tipping it onto a tissue as before.
4. Rinse slides well. For this purpose, apply TBS 1× using 500 μL/slide for 5 min. Repeat the whole process 3 times (*see* **Note 11**).

3.4 Secondary Incubation

1. Prepare 10% NGS in 4% TBS-T according to the recipe above using 500 μL/slide. Then, add the secondary antibody of choice to the prepared buffer. For example, if the primary antibody was raised in a rabbit, the secondary antibody must be an anti-rabbit secondary (*see* **Note 12**).
2. Then, apply the prepared buffer to the slides, cover the chamber, and set aside for 2 h (*see* **Note 13**).
3. After this time, remove the secondary antibody solution, tipping it onto a tissue as before.
4. Rinse slides well. For this purpose, apply TBS 1× using 500 μL/slide for 5 min. Repeat the whole process 3 times.

3.5 TrueBlack® Incubation

1. Apply TrueBlack® prepared according to the manufacturer's instructions (a 20× dilution of stock solution in 70% ethanol) to the slide using 500 μL/slide, then cover the slides and wait 5 min (*see* **Note 14**).
2. Rinse slides well. For this purpose, apply TBS 1× using 500 μL/slide for 5 min. Repeat the whole process 3 times (*see* **Note 15**).

3.6 Coverslip Slides

1. Dry the edges of the slide with a paper towel, apply 2 drops of Fluoromount-G containing DAPI using a plastic Pasteur pipette, and place a coverslip on top (*see* **Note 16**).
2. Store the prepared slides in a cardboard slide holder at a temperature of 4 °C (*see* **Note 17**).

4 Notes

1. The concentration of TBS-T may vary depending on both the analyzed tissue and the thickness of the prepared sections. The given concentration of 4% TBS-T is optimal for slide-mounted

40 μm sections of murine histological preparations such as brain, liver, muscle, and bowel. The buffer mixing time can be extended depending on your needs. The buffer can be prepared in larger quantities and stored at 4 °C for up to several months.

2. The assumed volume of solution for one slide is 500 μL; therefore, the number of slides prepared for staining should be multiplied by 500 to give the total volume of buffer needed in μL. We prepare the blocking buffer immediately before staining. Additionally, it is possible to use Normal Serum from other species (typically the species in which the secondary antibody was raised), but then the same serum must be consistently used throughout the staining process. In our experience, the best results are obtained using normal serum from goat, rabbit, or donkey.
3. The amount of buffer depends on the quantity of slides. Please remember to consistently use the same buffer that was used to prepare the blocking buffer.
4. The purpose of using TrueBlack® is to mask the autofluorescence of the storage material that accumulates in lysosomal diseases and tissue histofluorescence as a result of aldehyde fixation of tissues. The amount of prepared buffer depends on the number of slides. TrueBlack® is a toxic substance; therefore, special care should be taken when working with this reagent.
5. Most reagents are light sensitive, so it is good practice to stain using a staining chamber with a well-fitting lid.
6. Applying a hydrophobic barrier using a PAP pen is not a mandatory step, but it facilitates the process of applying and maintaining the reagent on the section. It is good practice to outline the two long edges of the slide and clearly separate the colored end (where information is written) from the area containing the sections. This will help prevent this information from being erased during the staining procedure.
7. It is recommended to prepare an excess volume of buffer. It is often necessary to apply more than 500 μL of buffer per slide, and it is important to assess whether all sections are completely covered by buffer. The estimated time is 1 h, but it can be extended or shortened. It is important to be consistent throughout the staining process.
8. Placing the slide vertically, touch one edge of the slide to a tissue wipe so that the buffer flows freely off the slide.
9. The concentration of the antibody may vary depending on the antigen and tissue type, or between manufacturers, so it is good

practice to always perform test staining first to determine the optimal concentration.

10. Primary incubation should not be shorter than 2 h, but it is possible to extend this time. Alternatively, this step can be run overnight in a cold room, but care must be taken so that solutions do not evaporate, as this causes high levels of tissue fluorescence.
11. We can extend this time to 10 min for one rinse, but it is important to be consistent.
12. Fluorescent secondary antibodies are characterized by different color fluorochromes; therefore, when performing multi-channel staining, it is important to use antibodies that fluorescence in different channels. The concentration also depends on the brand of antibody used. Secondary antibodies are light-sensitive, so it is important to keep them and any buffer containing them in the dark.
13. You can modify the time to 1 h or even leave it longer up to 3 h; however, as always, it is important to be consistent.
14. Time depends on the thickness and structure of the tissue; usually, it is between 3 and 5 min for 40 μm thick sections. TrueBlack® is toxic and should therefore be disposed of in a biohazard waste container.
15. This time can be extended to 10 min for one rinse; the number of rinses may also vary. Sometimes two rinses are enough.
16. The use of Fluoromount-G containing DAPI makes cell nuclei visible. This is useful when scanning slides for analysis, as the slide scanner can be set to focus on the DAPI channel, preventing fading of any of the channels containing antigen-specific signals.
17. Over time, the quality of immunofluorescent staining may deteriorate; therefore, it is recommended to analyze it within the next 6 months.

5 Conclusions

The basic method outlined above has been used successfully in our laboratory to stain CNS tissue and preparations of bowel, muscle, and liver tissue from many different species (Figs. 1, 2, and 3). Following careful optimization, this method provides consistent staining of a broad range of antigens with minimal background, minimizing variation and ensuring consistency of data within different treatment groups and genotypes. This methodology also provides superior sensitivity over traditional free-floating

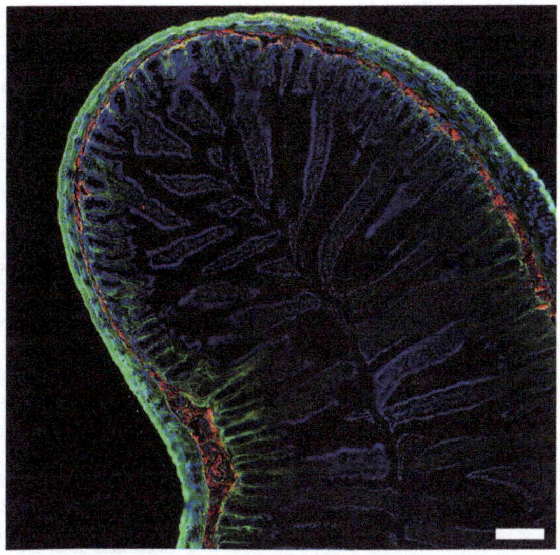

Fig. 1 Example staining of a cross-section of the bowel of a wild-type mouse. Staining was performed using NGS with primary rabbit anti-laminin (green), rat anti-Sox2 (red), and DAPI (blue). TrueBlack® counterstain was used to eliminate background

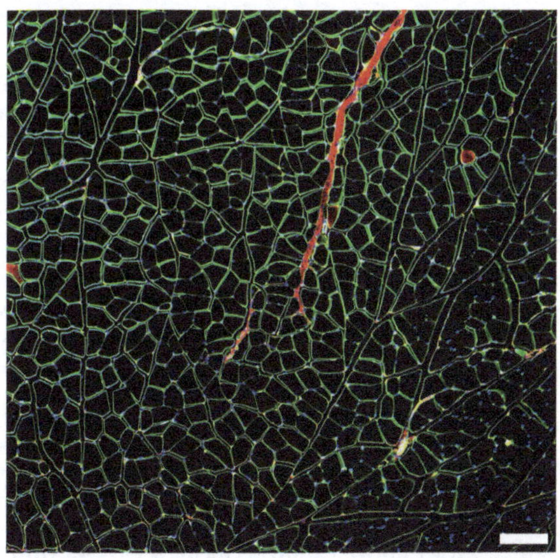

Fig. 2 Example staining of a cross-section of the quadriceps femoris muscle of a wild-type mouse. Staining was performed using NGS along with primary rabbit anti-laminin (green), rat anti-CD31 (red), and DAPI (blue). TrueBlack® counterstain was used to eliminate background

immunoperoxidase methods and permits the simultaneous visualization of multiple antigens. We have published data of this type in many studies, and specific protocols for different antigens are available upon request.

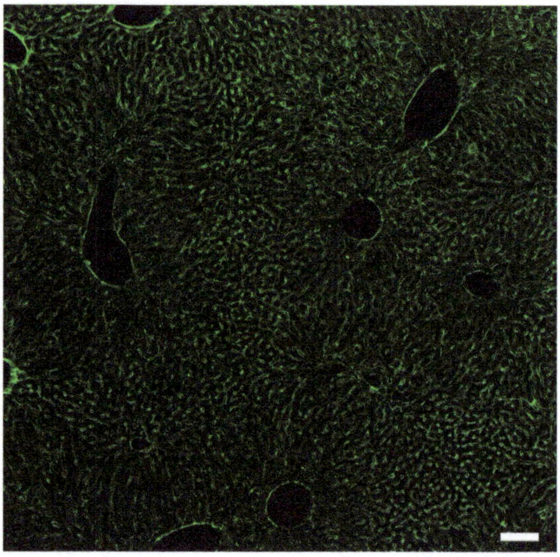

Fig. 3 Example staining of a liver section from a CLN2 AAV9-treated mouse. Staining was performed using NGS with a rabbit anti-TPP1 primary antibody (green) to detect AAV9-driven expression of TPP1. TrueBlack® counterstain was used to eliminate the background

Acknowledgements

The studies described in this chapter were supported by grants to JDC from Noah's Hope—Hope 4 Bridget, Fore Batten Foundation, the Children's Brain Diseases Foundation, Beyond Batten Disease Foundation, The Lehrman Family Fund, and Haley's Heroes, as well as by institutional support from the Department of Pediatrics, Washington University in St. Louis. Additional support was provided by NINDS National Institutes of Health grants R21 NS116574, R21 NS126907, and R01 NS124655 to JDC. We also thank Dr. Alison Barnwell for constructive comments on the manuscript.

References

1. Koivunen ME, Krogsrud RL (2006) Principles of immunochemical techniques used in clinical laboratories. Labmedicine 37:490–497
2. Magaki S, Hojat SA, Wei B, So A, Yong WH, Yong WH (2019) An introduction to the performance of immunohistochemistry. Methods Mol Biol 1897:289
3. Li A, Yang D-H, Vancurova I, Zhu Y (2020) Application of immunohistochemistry in basic and clinical studies. Methods Mol Biol 2108: 43–55
4. Moreno V, Smith EA, Piña-Oviedo S (2022) Fluorescent immunohistochemistry. Methods Mol Biol 2422:131–146
5. Kruger NJ (2003) Detection of polypeptides on immunoblots using secondary antibodies or protein A. Methods Mol Biol 32:215
6. Im K, Mareninov S, Diaz MFP, Yong WH (2019) An introduction to performing immunofluorescence staining. Methods Mol Biol 1897:299–311
7. Hofman F (2002) Immunohistochemistry. Curr Protoc Immunol Chapter 21:4

8. Coons AH, Creech HJ, Jones RN (1941) Immunological properties of an antibody containing a fluorescent group. Proc Soc Exp Biol Med 47(2):200–202
9. Swanson PE (1988) Foundations of immunohistochemistry. A practical review. Am J Clin Pathol 90(3):333–339
10. DeLellis RA, Sternberger LA, Mann RB, Banks PM, Nakane PK (1979) Immunoperoxidase technics in diagnostic pathology. Report of a workshop sponsored by the National Cancer Institute. Am J Clin Pathol 71(5):483–488
11. Mesa-Tejada R, Pascal R, Fenoglio CM (1977) Immunoperoxidase: a sensitive immunohistochemical technique as a "special stain" in the diagnostic pathology laboratory. Hum Pathol 8(3):313
12. Ortiz Hidalgo C (2022) Immunohistochemistry in historical perspective: knowing the past to understand the present. Methods Mol Biol 2422:17–31
13. Elias JM (1999) Immunohistochemistry: a brief historical perspective. J Histotechnol 22(3):163–167
14. Aerts JMFG, Kallemeijn WW, Wegdam W, Ferraz MJ, van Breemen MJ, Dekker N, Kramer GJ, Poorthuis BJHM, Groener JEM, Cox-Brinkman J, Rombach SM, Hollak CEM, Linthorst GE, Witte MD, Gold H, van der Marel GA, Overkleeft HS, Boot RG (2011) Biomarkers in the diagnosis of lysosomal storage disorders: proteins, lipids, and inhibodies. J Inherit Metab Dis 34(3):605–619
15. Parkinson-Lawrence E, Fuller M, Hopwood JJ, Meikle PJ, Brooks DA (2006) Immunochemistry of lysosomal storage disorders. Clin Chem 52(9):1660–1668

Index

Stephanie M. Hughes and Indranil Basak (eds.), *Lysosomes: Methods and Protocols*, Methods in Molecular Biology, vol. 2976, https://doi.org/10.1007/978-1-0716-4844-5,